Rethinking Randomness

A New Foundation for

Stochastic Modeling

Jeffrey P. Buzen

Library of Congress Control Number: 2015910498
CreateSpace Independent Publishing Platform, North Charleston, SC

CreateSpace, an Amazon.com Company

Printed in the United States

Cover illustration by Roger Burrows
Cover and book design by Jeffrey P. Buzen

Library of Congress Cataloging-in-Publication Data

Buzen, Jeffrey P. 1943 –
Rethinking randomness: a new foundation for stochastic modeling/
Jeffrey P. Buzen
330 pages   15.24 × 22.86 cm
Includes bibliographic references and index.
ISBN: 978-1-508-43598-3 (pbk.)
1. Probability. 2. Markov chains. 3. Stochastic processes. 4. Queuing models.
I. Title
2015910498

First Edition – August 2015

This book is dedicated to my wife Judy
whose consistent caring and support
have been of immeasurable value
during the writing of this book and
throughout the decades of work that preceded it.

About the Cover

The illustration on the cover of this book appears with the permission of its creator, Roger Burrows. This image is the result of a multi-step process. The first step employs algorithms that arrange spheres of different sizes to form stable, closely packed three-dimensional structures. Slicing these structures with flat planes exposes two dimensional cross sections of the type shown on the upper half of the next page.

The next step involves transforming these cross sections into lattices by connecting pairs of circle-to-circle contact points with straight lines. A partially completed lattice is shown on the lower half of the next page. For more information about close packing algorithms and lattices, see "Making Space – Finding Space" by Roger Burrows and the website www.rogerburrowsimages.com.

The final step involves artistic creativity rather than algorithmic precision. By selectively coloring individual segments of the complete lattice and adding a few artistic flourishes, it is possible to create the cover's whimsical undersea panorama. Selective coloring processes of this type form the basis for Roger Burrows' series of *Ultimate Coloring Experience* books: *Images 1*, *2*, *3*, *4* and *5*. See also his *Hidden Images* series and his other publications.

Note that the discovery of images that are hidden within patterns generated by geometric algorithms is similar in spirit to the discovery of algebraic relationships that are hidden within numerical data generated by random processes. Hidden algebraic relationships of this type play a crucial role in observational stochastics. Many of the results and examples presented in this book are based on such relationships.

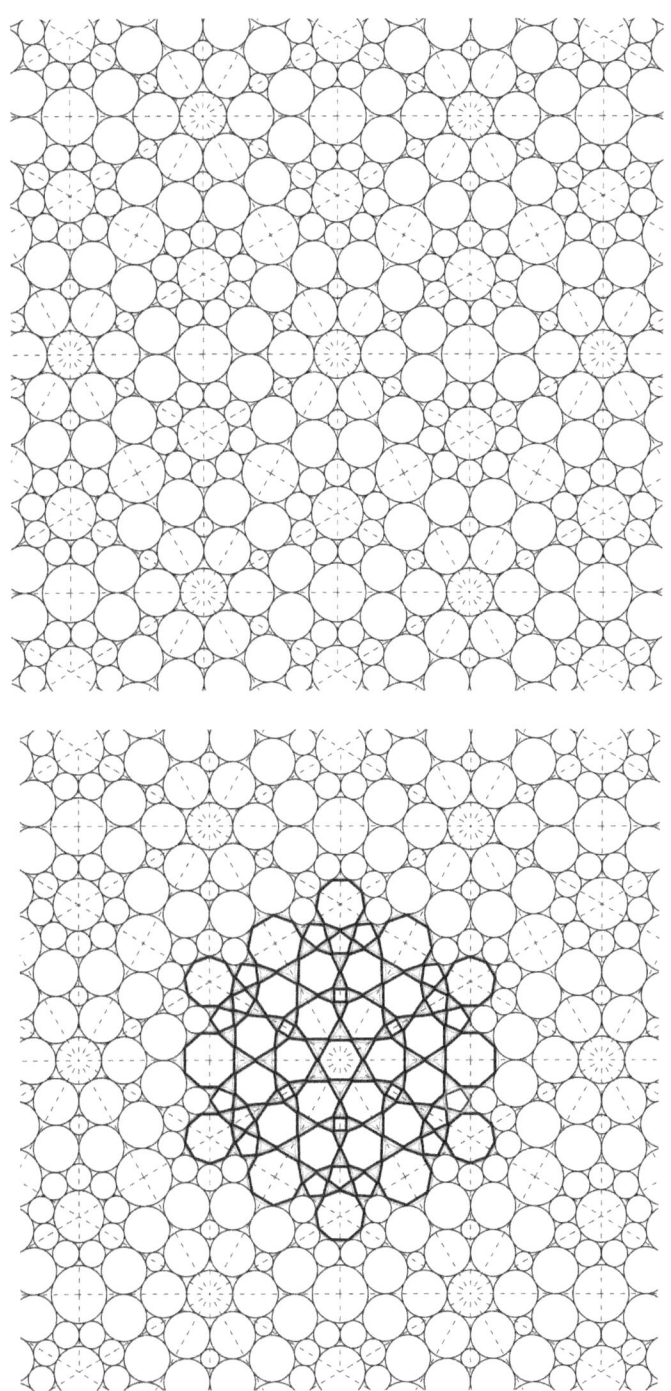

Preface

Rethinking Randomness presents my thoughts on a puzzle I have been wrestling with for many years. In the early 1970s I developed a mathematical model that could predict the utilization levels, throughput rates and response times of large mainframe computer systems. The central server model, which is described briefly in Section 7.11.2, formed the basis of my PhD dissertation in Applied Mathematics at Harvard. As a theoretician, I was most interested in the challenge of creating realistic yet solvable analytic models and in developing algorithms that made their evaluation computationally feasible (Buzen's algorithm: Wikipedia).

Being the first of its type, the central server model quickly attracted the attention of practitioners who applied it to questions of importance within their own datacenters. As part of their investigations, these analysts validated their results by comparing predicted values of utilization, throughput and response time with actual measured values. The model proved remarkably accurate in practice.

I was delighted to learn of these successful validations. However, I was also puzzled by the fact that real world computer systems did not appear to satisfy the stringent mathematical assumptions that traditional queuing models require. This piqued my curiosity: why did these models exhibit such an unexpectedly high degree of accuracy? I have spent much of the past four decades investigating this issue and identifying conditions under which probabilistic models do, and do not, work well in practice.

My first paper on this topic, "Fundamental Laws of Computer System Performance," was published in 1976. It introduced the concept of operational analysis, an approach that mirrors the operational procedures practitioners follow when they apply certain mathematical equations to the analysis of real world systems. This paper received the inaugural "Test of Time" Award from ACM Sigmetrics in 2010, 34 years after its original date of publication.

Operational analysis has been integrated into a succession of major texts dealing with the analysis of computer performance (Ferrari, Serazzi and Zeigner 1983), (Lazowska, Zahorjan, Graham and Sevcik 1984), (Jain 1991), (Menascé, Almeida and Dowdy 1994), (Menascé and Almeida 2001), (Gelenbe and Mitrani 2010), (Harchol-Balter 2013), (Denning and Martell 2015). Nevertheless, it remains relatively unknown among engineers, experimental scientists and other practitioners who apply probabilistic models to problems in the physical, social and life sciences. A primary goal of this book is to recast the basic principles of operational analysis in a more general context so that its core ideas become accessible to this wider community.

Rethinking Randomness presents an introduction to observational stochastics, the successor to operational analysis. Observational stochastics extends the mathematical foundation of the original theory in several ways: most importantly, by formalizing the intuitive notion of observing and measuring the behavior of a system as it operates during an interval of time. Students enrolled in courses on probability and stochastic modeling should find observational stochastics particularly helpful in understanding how the material they are studying in class is actually applied in practice. And because all mathematical arguments are self contained and relatively straightforward, technically oriented non-specialists who wish to explore the connection between probability theory and the physical world should find much of the material in this book readily accessible.

I would like to thank my colleague Yaakov Kogan, formerly of Bell Labs and AT&T, for suggestions that have improved the clarity and strengthened the mathematical rigor of this book.

Jeffrey P. Buzen

Note: Corrections, clarifications and supplementary information will be available at www.RethinkingRandomness.com.

Contents

5 Modeling Techniques and Examples.......... 103

6 Simple Continuous Time Models.................... 143

CHAPTER 1

Overview of Observational Stochastics

1.1 Mathematics, measurement and the physical world

Physical objects and processes have existed in one form or another since the birth of the universe. They are real entities whose existence does not depend in any way on the mathematical models that characterize their properties and behavior. In contrast, mathematical models are pure creations of the human mind. They exist in an abstract realm where they can be defined and analyzed without referring to physical entities of any type.

When mathematical models are applied to practical problems, the symbolic variables that appear in their associated equations are typically interpreted as measurable properties of corresponding physical entities. In effect, the process of observing and measuring physical entities provides a crucial link that connects the realm of pure mathematics with physical reality. This linkage is immediately apparent in the work of engineers and experimental scientists who use mathematical models to solve problems that arise in the real world.

The connection between models and reality is not always so direct. One especially important exception arises when mathematical models are used to analyze processes whose detailed step-by-step behavior follows irregular patterns that appear to be the product of random factors. In such cases, the observable properties of these processes are traditionally regarded as samples that have been drawn from underlying probability distributions. The parameters that characterize these distributions (e.g., means, variances, etc.) appear as symbolic variables in the equations derived from such models.

Distributional parameters of this type differ fundamentally from the symbolic variables employed in conventional mathematical models. They are not linked directly to the observable properties of individual physical entities. At best, their exact values can be estimated with varying degrees of confidence to lie within ranges that are specified by sets of upper and lower bounds.

This book develops an alternative framework for analyzing systems and processes whose behavior appears to be driven by random (i.e., non-deterministic) forces. The new approach is based on a simple principle that is employed routinely in many other scientific disciplines: all symbolic variables that appear in the specification and solution of a mathematical model must represent directly observable properties of the system or process being modeled. To highlight this point, the alternative framework will be referred to as *observational stochastics*.

In effect, observational stochastics complements traditional stochastic modeling by analyzing what are essentially the same physical phenomena from the perspective of individuals who apply mathematical models (i.e., practitioners) rather than theorists who work primarily in the realm of pure mathematics.. This alternative path leads to vastly simpler derivations of equations that are analogs of major results from stochastic theory, a different foundation for conceptualizing the intuitive notion of randomness and uncertainty (linked in spirit to nineteenth rather than twentieth century mathematics), new tests for determining whether or not a model's assumptions are actually satisfied in practice, and new procedures for bounding errors when a model's assumptions are not satisfied exactly.

1.2 Highlights of observational stochastics

Observational stochastics formalizes the procedures that practitioners follow when they validate mathematical models based on stationary (i.e., steady state) stochastic processes.   These validation procedures typically begin by observing the behavior of an actual system during a finite interval of time. The objective is to determine whether or not the system's behavior conforms to equations derived from the model being validated.

To make this determination, all symbolic variables that appear in these equations are associated with directly observable quantities. These quantities are then measured so the equations can be evaluated numerically.   If the resulting computations generate predicted values that agree with their directly measured counterparts, the validation is a success.

Observational stochastics employs several interrelated concepts to formalize the process of model validation. The next few sections provide brief descriptions of these concepts.   More detailed descriptions, along with numerous examples, are presented in the remaining sections of Chapter 1 and in the chapters that follow.

1.2.1 Interval-wide proportions

When analyzing a traditional stochastic model of a real or hypothetical system, one of the primary objectives is to derive the steady state probability distribution of the underlying stochastic process.   This distribution represents the chance that the system being modeled is in a given state at some instant of time.

While this characterization is meaningful within the realm of pure mathematics, the state of an actual system at any instant during a traditional validation experiment is not a distribution. It is instead a specific observable value. This apparent mismatch is resolved by reinterpreting   instantaneous   probabilities   as   interval-wide proportions. In other words, the steady state probability that the system is in some given state is equated to the proportion of time the system spends in that state during the entire validation interval.

The formal justification for this intuitively appealing assumption is based a subtle mathematical result, the Ergodic Theorem, which is discussed further in Section 2.7 and Section 9.2.

Observational stochastics avoids the reinterpretation issue entirely by dealing with observable interval-wide proportions from the onset. This shift in initial focus is one of the principal defining characteristics of observational stochastics.

1.2.2 Directly measurable variables

All symbolic variables employed in observational stochastics (including the interval-wide proportions mentioned in the preceding section) represent directly measurable quantities. These variables are typically defined as ratios of raw counts and durations measured over entire intervals or over subsets of entire intervals. Section 3.7 introduces a formal modeling framework within which these directly observable variables can be defined. Straightforward procedures for measuring the values these variables attain in each specific instance are specified in Sections 3.7.3 and 3.7.4.

1.2.3 Directly verifiable assumptions

The assumptions employed in observational stochastics are expressed as algebraic relationships among symbolic variables that represent directly observable quantities. To verify a particular assumption, the numerical values that these variables actually attain can be measured and then tested to determine if the corresponding algebraic relationship is satisfied. If so, it is possible to conclude with complete certainty that the assumption is satisfied for the interval in question.

Verification procedures of this type are employed routinely in science and engineering. However, because of the way randomness is represented in traditional probabilistic models, the assumptions incorporated into these models cannot be verified with complete certainty through straightforward procedures of this type. The next section examines why this is so.

1.2.4 Sampling premise excluded

Mathematicians who deal with uncertainty typically associate the intuitive notion of randomness with the process of drawing samples from underlying probability distributions. The assumption that observed values can be regarded as samples drawn from such distributions will be referred to as the *sampling premise*.   In addition to its powerful intuitive appeal, the sampling premise provides the foundation for most – if not all – traditional applications of probability and statistics.

Despite its nearly universal acceptance, the sampling premise does not represent a verifiable hypothesis. An unlimited number of different probability distributions can generate exactly the same set of observed values when a finite number of samples are drawn at random from these distributions. Thus the validity of the sampling premise (for a given distributional form and a specific set of parameter values) can never be established with complete certainty by inspecting sets of observed values. Since observational stochastics requires directly verifiable assumptions, an alternative to the sampling premise must be introduced.

1.2.5 Loose constraints versus distributional assumptions

The primary role of the sampling premise is to provide a platform for specifying assumptions about the nature of the distributions incorporated into individual models. For example, the lengths of messages flowing through a communications network are often modeled as samples drawn from exponential distributions (Section 6.3). Distributional assumptions of this type enable analysts to characterize the most important aspects of a process's behavior, while allowing the step-by-step details to remain uncertain.

The new class of assumptions that are used within observational stochastics as replacements for the sampling premise are based on the simple idea of *loose constraints*. Essentially, loose constraints are algebraic relationships among symbolic variables that represent directly observable quantities and serve as a model's parameters (see Section 5.1.3).

The symbolic variables that appear in loose constraints correspond to interval-wide proportions and averages. Individual observable values that are combined together to form these interval-wide quantities have no direct bearing on any results that are ultimately derived. These individual observable values represent immaterial details whose exact values can, in principle, remain uncertain.

To illustrate the form loose constraints typically take, suppose a coin is tossed 1000 times. Suppose further that 500 of these tosses actually come up heads. Thus p, the observed proportion of heads, is equal to one half.

Now consider only those tosses that follow immediately after one of the 500 tosses that came up heads. A typical loose constraint would require that the proportion of tosses that come up heads in this subset must also be equal to one half. More generally, a loose constraint might require that p, the overall proportion of heads observed for all tosses, must be equal to p-sub, the proportion of heads observed in some well defined subset of tosses.

Note that loose constraints of this type are directly verifiable. They also allow immaterial step-by step details to remain uncertain. There is no need to assume that the outcome of each coin toss corresponds to a sample drawn from a probability distribution. The sampling premise is never invoked. In obscr- vational stochastics, it is sufficient to base an analysis entirely on relationships among directly observable quantities such as p and p-sub. In essence, observational stochastics replaces assumptions about the way observable values have been generated (i.e., the sampling premise) with assumptions about the way these values actually appear (i.e., assumptions expressed as loose constraints).

Many major results from the traditional theory of Markov processes, and from other branches of stochastic modeling, have direct counterparts that can be derived using such assumptions. In addition to providing a simpler and more intuitive framework for their derivation, observational stochastics also demonstrates that these results are valid under conditions that are substantially more general than is commonly recognized.

Note that the mathematical relationships derived through observational stochastics apply to all trajectories that satisfy the loose constraints of the corresponding model. This is analogous to the mathematical relationships derived through traditional stochastic models, which apply almost surely to the ensemble of sample paths associated with the underlying stochastic process. These points are discussed further in Sections 6.9 and 9.2.

In contrast, the mathematical relationships derived using a conventional deterministic model apply only to a single trajectory whose detailed step-by-step structure is fully determined by the parameters of the deterministic model being analyzed. This distinguishes deterministic modeling from both observational stochastics and traditional stochastic analysis.

1.2.6 Loosely constrained deterministic (LCD) models

Validation experiments always involve two distinct entities: an abstract mathematical model and a physical system that can be observed and measured during an interval of time. In traditional stochastic analyses, the abstract mathematical model is a stochastic process: that is, an ordered sequence of random variables indexed by integers or by real numbers. The ordering represents the flow of time, and the probability distribution associated with each random variable in the sequence represents the distribution of states of the stochastic process at the corresponding instant.

Observational stochastics is based on a significantly different mathematical abstraction referred to as a *loosely constrained deterministic (LCD) model* (Buzen 2012). LCD models reflect the fact that system behavior is typically controlled by a combination of deterministic and uncertain factors. For example, when a new customer arrives at a queue, the length of the queue increases by one. This increase is entirely deterministic. However, the elapsed time between successive arrivals at a queue is typically uncertain.

Deterministic aspects of system behavior are represented within LCD models by formalisms adapted from Computer Science and Electrical Engineering: specifically, finite state automata (Mealy 1955). Loose constraints on the workloads processed by these automata and on the trajectories these workloads generate enable detailed aspects of system behavior to remain uncertain.

LCD models play a crucial role in observational stochastics by providing a rigorous foundation for the definitions of observability, measurability and system state. Section 3.7 develops a formal characterization of LCD models and related concepts.

1.2.7 Finite scope

Most of the derivations presented in this book are surprisingly simple and intuitive. The corresponding derivations in traditional stochastic analysis make use of concepts with complex definitions involving the limiting properties of infinite sequences.

The simple and intuitive nature of observational stochastics stems from its direct connection to the validation paradigm described at the start of Section 1.2. This connection links the symbolic variables of observational stochastics to observable properties of entities that exist in the real world. The powerful theorems discussed in Chapter 9 are never required to bridge the gap between real world observables and the formal mathematical abstractions encountered within the traditional theory of stochastic processes.

An equally important but less obvious point is that the validation paradigm is essentially finite in nature. In observational stochastics, intervals are always finite, as are the number of state-to-state transitions that can be observed during such intervals. These restrictions, which are discussed further in Section 8.6 and in the Epilog, should have no impact on the day-to-day concerns of most practitioners. As noted by Kolmogorov (1933), the creator of the modern axiomatic theory of probability, "In describing any observable random process, we can obtain only finite fields of probability. Infinite fields of probability occur only as idealized models of real random processes."

Chapter 9 includes a very brief introduction to the measure theoretic concepts required to deal with cases where the number of state-to-state transitions can become infinite. In such cases, the entirely straightforward measurement procedures specified in Chapter 3 must be replaced by a more general mechanism known as Borel measure (or its extension, Lebesgue measure). As discussed in Chapter 9, this would add a substantial amount of complexity to the mathematical analyses presented here, while providing little of practical value.

Note that omitting measure theoretic extensions from the formal specification of observational stochastics does not reduce its level of mathematical rigor. It merely restricts the applicability of derived results to cases that fall within the finite boundaries of the validation paradigm. For most practitioners, these are the only cases that really matter.

Table 1-1 summarizes the principal characteristics of observational stochastics and contrasts them with their traditional stochastic counterparts.

	OBSERVATIONAL STOCHASTICS	TRADITIONAL STOCHASTIC MODELS
In typical cases, the primary objective of an analysis is to derive expressions for:	Trajectory-based proportions	Probability distributions, which may be stationary (steady state) or transient (time dependent)
Symbolic variables used in modeling equations represent:	Trajectory-based quantities that are directly measurable	Distributional parameters + stationary or transient probability values
Modeling assumptions are expressed in terms of:	Directly observable properties of trajectories (loose constraints)	Mathematical properties of the associated stochastic process
Directly observable values of inherently uncertain quantities (e.g., results of a sequence of coin tosses) are regarded as:	Immaterial details whose aggregate values are subject to explicitly specified loose constraints	Samples drawn at random from explicitly specified probability distributions (referred to in Section 1.2.4 as the *sampling premise*)
The formal mathematical objects at the center of the analysis are:	Loosely constrained deterministic models (i.e., LCD models)	Stochastic processes (i.e., sequences of random variables)
The length of a modeling interval is:	Always finite (which meets the needs of most practitioners)	Either finite or infinite

Table 1-1.

Observational stochastics versus traditional stochastic modeling

1.2.8 Operational analysis

Observational stochastics represents a refinement and extension of this author's earlier work on operational analysis (Buzen 1976a), (Buzen 1976b), (Denning and Buzen 1978). Both methods are based on the idea that symbolic variables should be defined in terms of the operational steps that practitioners follow when the values of these variables are actually measured. Both methods deal with interval-wide proportions and directly verifiable assumptions. However, operational analysis is built upon an intuitive informal foundation. There is no underlying LCD model. This makes it difficult to answer the most basic of questions: what exactly is a *system*, an *observable state*, and a *directly measurable quantity?* LCD models make it possible to provide precise mathematically rigorous answers to these questions and to resolve some unfortunate misunderstandings that have arisen regarding the nature and value of operational analysis.

Another significant refinement concerns the relationship between uncertainty, immateriality and loose constraints. These concepts, which were not addressed explicitly in operational analysis, are central elements in observational stochastics and represent an alternative perspective on the characterization and analysis of randomness.

Despite these and other differences, it is still true that many results in observational stochastics are derived using the same assumptions and step-by-step arguments that were originally employed in operational analysis. References to this earlier work appear in subsequent chapters.

1.3 A simple deterministic example

Observational stochastics has much in common with the approach to modeling that is employed in many branches of science and engineering. To illustrate this point, consider a simple example: suppose a heavy ball is tossed straight up into the air. Under the standard assumptions of Newtonian mechanics, the ball will decelerate steadily until it reaches some maximum altitude and then fall back to earth.

Suppose an observer can measure the initial speed of the ball and the maximum height it reaches. Such an observer may wish to determine the relationship between these two directly observable quantities in order to predict how much higher the ball would go if its initial velocity were doubled.

This problem can be solved by applying mathematical models found in most introductory texts on classical physics. The exact form of the solution is not of concern here. Instead, the important point is that the ball's maximum height can be predicted using a simple equation that depends only on the ball's initial velocity and the gravitational constant g. The analysis that leads to this solution can be extended to calculate the ball's exact height at each instant during its flight. This relationship is depicted in Figure 1-1.

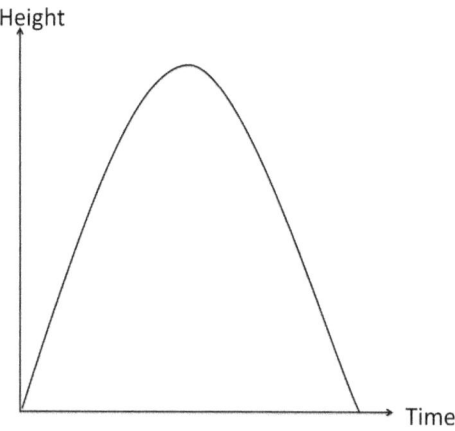

Figure 1-1. Trajectory of a tossed ball

Curves of the type displayed in Figure 1-1 will be referred to as trajectories. Note that this terminology differs slightly from standard usage because the horizontal axis represents elapsed time rather than distance traveled along the ground.

The equation that specifies the trajectory in Figure 1-1 can be used to determine the proportion of time the ball spends at various altitudes. For example, this equation can be used to determine the following quantities:

$P(0)$ = proportion of time the ball is at or below 25% of its maximum altitude.

$P(1)$ = proportion of time the ball is above 25%, but at or below 50%, of its maximum altitude.

$P(2)$ = proportion of time the ball is above 50%, but at or below 75%, of its maximum altitude

$P(3)$ = proportion of time the ball is above 75% of its maximum altitude.

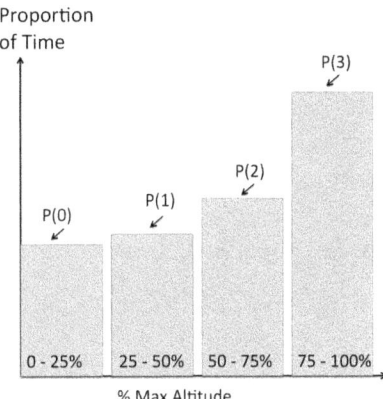

Figure 1-2. Proportion of time at each altitude

1.4 A simple random walk

Random walks are a standard topic in many introductory texts on probability theory. They share several important properties with the deterministic example discussed in the preceding section. Imagine a walker travels along a route marked by four stations that are positioned along a straight line as shown in Figure 1-3. As the walk proceeds, the walker departs from the current station, turns left or right, and moves to the next station in the path.

Figure 1-3. Random walk with four stations

To prevent the walker from disappearing off the end of the path, assume there are reflecting barriers at each end. Thus a walker exiting from station 3 and turning right will encounter a reflecting barrier and be directed immediately back to station 3. Similarly, a walker exiting from station 0 and turning left will encounter a reflecting barrier that directs the walker back to station 0.

The sequence of stations the walker passes through can be displayed in a simple diagram as shown in Figure 1-4. The horizontal axis represents time, just as it does in Figure 1-1. However, in this case time is discrete rather than continuous. Each dot represents an instant at which the walker departs from the current station and moves to the next. Note that the next station is identical to the current station if the walker encounters a reflecting barrier. The dots are connected by a continuous line to make it easier to visualize the walker's path.

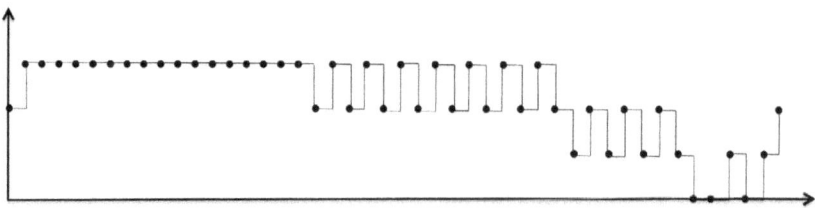

Figure 1-4. Trajectory of a walker

The vertical axis in Figure 1-4 contains only four discrete points (corresponding to stations 0, 1, 2 and 3. In contrast, the vertical axis in Figure 1-1 represents the height of the ball, which is of course a continuous variable.

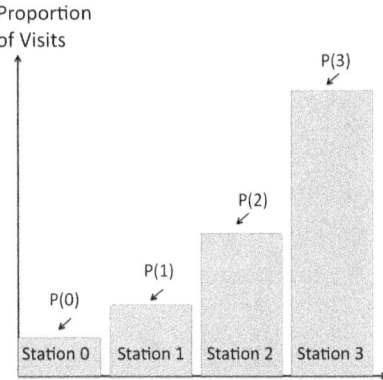

Figure 1-5. Proportion of visits to each station

Despite the discrete nature of the horizontal and vertical axes, it remains useful to regard Figure 1-4 as a trajectory that depicts the step-by-step changes in the location of the walker as a function of time. As in the case of a ball tossed into the air, it is also possible to consider the proportion of visits that the walker makes to each station over the course of the trajectory. Such a distribution might have the form illustrated in Figure 1-5.

1.4.1 Random walk: solution

Suppose an analyst wishes to predict the proportions that appear in Figure 1-5. To make these predictions, it is necessary to provide some additional information about the behavior of the walker. One obviously important factor is the proportion of all turns that are made to the right (rather than the left) over the course of the trajectory. Assume that this proportion is represented by the letter R. The value of R is a directly observable property of any trajectory of the type illustrated in Figure 1-4.

By itself, the value of R is not sufficient to determine the desired proportions. It is thus necessary to introduce additional assumptions that preserve the essential unpredictability and uncertainty of the random walk while also providing enough mathematical structure to obtain a solution.

It is traditional to assume at this point that the walker's trajectory is driven by a random process such as tossing a coin, rolling a pair of dice, or drawing a card from a deck. As discussed in detail in Chapter 2, this assumption leads to a stochastic model that ultimately yields the following predictions for P(n), the observed proportion of visits the walker makes to station n:

$$P(0) = \frac{1}{1 + R/(1-R) + [R/(1-R)]^2 + [R/(1-R)]^3} \tag{1-1}$$

$$P(1) = P(0) \times R/(1-R) \tag{1-2}$$

$$P(2) = P(0) \times [R/(1-R)]^2 \tag{1-3}$$

$$P(3) = P(0) \times [R/(1-R)]^3 \tag{1-4}$$

Experiments conducted by tossing coins or rolling dice are likely to agree closely with these predictions, provided the walk is allowed to continue for a large number of steps.

1.4.2 Probabilistic and observable variables

The solution presented in equations (1-1) through (1-4) is based on a stochastic model in which the direction of each turn is specified by a probability distribution: r is the probability that the next turn is to the right, and 1-r is the probability that the next turn is to the left. Note that r is a distributional parameter that does not correspond to a directly observable property of an actual walk.

In contrast, the proportion R that appears in equations (1-1) through (1-4) is directly observable. Its exact value can be expressed as a function of measurable properties of the trajectory in Figure 1-4.

For the practitioner, this subtle distinction is of no concern. Practitioners routinely interpret probabilities as observable proportions. As already noted, this interpretation can be justified through sound mathematical arguments. However, these arguments are only applicable if the physical process being modeled (in this case, the random walk) conforms exactly to the probabilistic assumptions of the associated stochastic model. Observational stochastics provides an alternative modeling framework that employs readily verifiable assumptions and is based entirely on directly observable variables such as $P(0)$, $P(1)$, $P(2)$, $P(3)$ and R.

1.4.3 Trajectories in observational stochastics

Trajectories of the type illustrated in Figure 1-4 play a central role in observational stochastics. They provide the basis for defining exactly what it means for a quantity to be directly measureable. For example, the value of R can be obtained from Figure 1-4 by counting the number of right turns that occur during the course of the trajectory and dividing that value by the total number of turns. This simple and intuitive concept is formalized in Chapter 3.

The trajectories in Figures 1-1 and 1-4 have much in common. However, there is one obvious and fundamentally important difference. Figure 1-1 is associated with a deterministic process while Figure 1-4 is not.

In the case of Figure 1-1, one would expect to observe exactly the same trajectory each time the ball is tossed straight up into the air at the same initial velocity. On the other hand, phenomena such as the random walk depicted in Figure 1-4 are not expected to follow predictable patterns, and there is no reason to believe that a second walk, operating under the same initial conditions as the first, would generate exactly the same trajectory.

Note that the solution presented in equations (1-1) through (1-4) is not affected by this type of uncertainty. Each value of P(n) is expressed as a function of R. The value of R is an interval-wide quantity that is not sensitive to the exact order in which stations are visited. As discussed in Section 1.2.5, these details are immaterial and can remain uncertain.

1.4.4 Loose constraints for random walks

Equations (1-1) through (1-4) express the directly observable proportions P(n) as functions of the directly observable quantity R. These equations are valid for some, but not all, trajectories. It is clearly important for practitioners to understand the conditions under which the validity of these equations can be guaranteed. Observational stochastics satisfies this objective by providing sets of directly verifiable assumptions (loose constraints) that are sufficient to guarantee the correctness of such solutions.

For random walks, these sufficiency assumptions are surprisingly simple. They are based on a self-evident insight: if the path a walker follows is intrinsically random, the direction of each turn should not be influenced by the walker's current location.

To formalize this particular aspect of randomness, let R(n) represent the proportion of turns the walker makes to the right after exiting from station n. The variables R(0), R(1), R(2) and R(3) represent observable proportions that can be extracted directly from trajectories of the type illustrated in Figure 1-4. If the direction of the next turn is not influenced by the walker's current location, all four proportions should be the same. This relationship among the values of R(n) is specified in equation (1-5).

$$R(0) = R(1) = R(2) = R(3) \hspace{4cm} (1\text{-}5)$$

Equation (1-5) is an example of a loose constraint. It expresses the intuitive notion that the direction of the next turn is independent of the walker's current location.   This form of independence is similar in spirit to statistical independence. However, within the framework of observational stochastics, the walker's current location and the direction of the next turn are not represented by random variables.   Thus, the formal definition of statistical independence is not applicable, even though the underlying concept is the same.   To avoid any misunderstanding, the term *empirical independence* will be used here to describe relationships of the type expressed by equation (1-5).

Chapter 3 presents a rigorous analysis of random walks within the framework of observational stochastics. The analysis demonstrates that equations (1-1) through (1-4) will accurately predict the values of P(n) for any trajectory that satisfies the assumption of empirical independence expressed by equation (1-5).

One minor technical assumption (matched endpoints) is also required to complete the proof of this somewhat surprising result. All other detailed properties of the trajectory are immaterial to the analysis and can remain uncertain.

1.4.5 Plausibility of empirical independence

A practitioner who wishes to apply equations (1-1) through (1-4) needs to know that the derivation of these equations is mathe-matically correct, and also that the assumptions upon which the derivation is based are likely to be satisfied during an actual random walk.

If a walker's turns are, in fact, determined by an inherently random physical process such as coin tossing, the random variable representing the direction of the next turn and the random variable representing the walker's current location will be statistically independent. Thus, it is entirely reasonable to expect empirical independence to be satisfied in such cases, especially if the trajectory is long.

Suppose, on the other hand, that the walker's path has not been generated by a physical process that can be characterized as intrinsically random. To imagine how such a situation might arise, suppose the walker is listening to music during the walk. If the walker happens to like the song being played, the next turn is to the right. If not, the next turn is to the left. Someone familiar with the walker's taste in music can predict the walker's trajectory from the songs on the playlist. This is clearly not a random process. However, an observer with no knowledge of the walker's strategy will simply notice an irregular sequence of left and right turns that follows no obvious pattern.

In this case it is entirely reasonable to assume that whether the walker likes or dislikes the next song will not be influenced in any way by the walker's current location. Equation (1-5) expresses the essence of this assumption in mathematical terms. There is no need to invoke the sampling premise and related concepts from probability theory to understand or apply equation (1-5). The intuitive rationale for this equation is based, not on the elusive and complex concept of randomness, but on the simple and directly observable notion of empirical independence.

1.4.6 Sensitivity analysis

Equation (1-5) states that all observed values of $R(n)$ are exactly the same. This constraint is unlikely to be satisfied precisely, even when there is strong reason to believe that the direction of the next turn is not influenced by the walker's current location. This is a legitimate concern for practitioners since it may reduce the accuracy of predictions made using equations (1-1) through (1-4).

Observational stochastics provides a direct method for dealing with this concern. As shown in Chapter 3, it is possible to relax the constraints imposed by equation (1-5) and derive a more general solution that expresses the values of P(n) as functions of the individual values of R(0), R(1), R(2) and R(3). This general solution, which again depends on the minor technical assumption of matched endpoints, will be exactly correct in all possible cases, whether or not the values of R(n) are all equal to each other.

This general solution makes it possible to carry out a sensitivity analysis to determine the maximum error that can arise when the observed values of R(n) are all within a specified percentage of the value of R. This information can be used by the practitioner to assess the risk of assuming empirical independence in cases where this assumption is not satisfied exactly. Such analyses have no direct counterparts in the traditional theory of stochastic processes. As discussed in Chapter 4, this type of sensitivity analysis can be extended even further by relaxing the technical assumption of matched endpoints.

1.5 Implications of observational stochastics

Observational stochastics provides a simple and intuitive framework for analyzing processes whose behavior appears to be driven by random forces. The link between uncertainty and immateriality is accommodated through the use of interval-wide variables. This allows the analysis to ignore immaterial details. More significantly, all required structural information is expressed in terms of directly verifiable loose constraints rather than conventional distributional assumptions. Thus the sampling premise is never required.

Observational stochastics supplements, but does not replace, conventional stochastic modeling. It provides direct benefits to practitioners and students while also raising interesting new problems for researchers and theoreticians to investigate.

1.5.1 For practitioners

Many practitioners deal with real systems whose behavior is irregular and unpredictable, but not necessarily driven by an intrinsically random mechanism such as coin tossing or random sampling. Observational stochastics legitimizes and formalizes the "back of the envelope" techniques that practitioners often employ in such cases.

Practitioners are, of course, most concerned with the accuracy of the stochastic models they employ in such cases. By identifying alternative sets of assumptions that are sufficient to ensure a model's accuracy, observational stochastics provides practitioners with a new way to assess the intuitive plausibility of the assumptions their conclusions depend on, new evidence-based tests to verify the validity of these assumptions, and new sensitivity analyses that can bound maximum errors when their model's assumptions are only approximately correct.

Observational stochastics also controls risk by limiting the number and scope of the predictions that can be derived from a given set of assumptions. The previous discussion of random walks illustrates this point.   As already noted, the assumption of empirical independence represented by equation (1-5) is sufficient to derive the solution expressed in equations (1-1) through (1-4). On the other hand, very few other properties of random walks can be derived using only the assumption represented by equation (1-5). For example, there is no way to determine the proportion of time that a walker exiting from station 0 will proceed directly to stations 1, 2 and 3 in succession by making three consecutive right turns. Equation (1-5) simply does not contain enough information to determine the proportion of time this sequence of turns will occur.

This question can still be analyzed within the framework of observational stochastics, but additional assumptions that characterize the walker's behavior at a finer level of detail must be introduced. Extensions of this sort are examined in Chapter 5.

In contrast, it is relatively easy to solve this particular problem using a classical stochastic model of a random walk. The classical model assumes that each turn is determined by sampling from a sequence of independent, identically distributed random variables. This assumption, which is substantially stronger than the assumption of empirical independence specified by equation (1-5), provides more than enough structure to answer the question.

In fact, the sampling assumption is strong enough to carry out analyses that are substantially more detailed: for example, determining the proportion of time a walker exiting from station 0 will cycle among stations 1, 2 and 3 at least one thousand times before finally returning to station 0. There is, in principle, no limit to the level of detail that can be extracted once a traditional stochastic model has been formulated.

While this capability may be quite appealing from a mathematical perspective, it is not ideal for practitioners because these exceptionally powerful assumptions make it difficult to identify the relative degree of risk inherent in the predictions made at each level of detail. In contrast, the comparative weakness of the assumptions employed in observational stochastics serves as a safety net. Analysts must explicitly introduce a succession of increasingly detailed assumptions until there is enough structure to solve the problem at hand. This incremental approach, which is discussed further in Chapter 5, makes it possible for practitioners to assess the intuitive plausibility and the risk of each additional layer of assumptions.

Once again, there is no simple counterpart to such incrementally detailed risk assessments in the traditional theory of stochastic processes. Traditional stochastic assumptions have an effectively limitless ability to generate results at ever increasing levels of detail. Some of these results will inevitably be riskier than others, but there is no easy way to tell since all are mathematically correct.

The simplicity and direct verifiability of observational assumptions can also prove useful to practitioners when seeking support for projects, products and services that involve stochastic models. Senior managers responsible for authorizing financial expenditures may not fully understand the powerful mathematical assumptions upon which traditional stochastic models are based. Framing the discussion in terms of observational stochastics can make it easier for these non-specialists to gain confidence in such models and, as a result, authorize appropriate levels of financial support.

1.5.2 For students

This book is also intended to serve as a companion text for students enrolled in introductory courses on probability theory or advanced courses on the theory and application of stochastic processes. Unlike conventional texts, there are no exercises at the end of each chapter. Instead, students are encouraged to take exercises from their own classroom texts, recast them within the framework of observational stochastics, and then consider how this change in perspective affects the assumptions required to obtain and apply solutions. In most cases, solutions will have the same algebraic form, but the manner in which observational solutions are obtained will involve fewer mathematical complexities. Also the solutions themselves will be linked more directly to practical applications. See Sections 5.7 and 8.2 for examples of this process.

Chapter 2 is intended primarily for students interested in traditional stochastic models of simple random walks. Certain subtle and often neglected issues are given special emphasis to highlight the interplay between theory and practice. These issues include the use of probabilistic equations to represent both the non-deterministic and the deterministic aspects of dynamic behavior, the nature and implications of steady state distributions, and the roles played by the Law of Large Numbers and the Ergodic Theorem when applying probabilistic results to problems that arise in the real world.

Throughout this book, important concepts will be introduced using simple illustrative examples that can be specified and analyzed with a minimum of mathematical notation. The conclusions drawn from these specific examples are subsequently extended to more general classes of problems. The primary goal is to strengthen the reader's insight and understanding rather than convey mathematical results in the most general and compact form.

Chapter 7 includes material for students and practitioners who have a specialized interest in queuing theory and its application to the analysis of computer systems and communication networks. Examples of some queuing models that have had an especially significant impact on the modern world are reviewed briefly.

Chapter 8 is a freestanding offshoot of the preceding seven chapters. Rather than dealing with problems that are traditionally analyzed using stochastic processes, this chapter deals instead with a broader class of questions that typically begin with the phrase "what is the chance".

The term *chance* is traditionally interpreted as *probability* when carrying out a formal analysis. However, *chance* can also be understood to represent an observable proportion. This interpretation is, of course, entirely consistent with the representation of uncertainty employed in observational stochastics. This proportionalist interpretation differs from both frequentist and Bayesian interpretations of probability; it is, however, closely aligned with Poincaré's classical view (Poincaré 1905, Chapter 11, "The Calculus of Probabilities"). Chapter 8 compares and contrasts these alternative views of probability, uncertainty and chance. Once again, the examples in Chapter 8 are intended to supplement the discussions in traditional introductory texts.

Chapter 9 extends the discussion in Chapter 8 by providing proportionalist treatments of the Law of Large Numbers and the Ergodic Theorem. As always, simple examples are used to provide intuitive insights into these theorems before more general and abstract mathematical arguments are presented.

1.5.3 For researchers

Observational stochastics is a work in progress. This book presents a number of topics that are promising candidates for further research. One such topic is shaped simulation, a technique discussed in Section 7.6. The essence of this technique can be illustrated using the familiar example of a random walk. Suppose a classical stochastic random walk is being analyzed by simulation. Suppose further that the sole objective of the simulation is to determine the values of $P(0)$, $P(1)$, $P(2)$ and $P(3)$. In such cases the simulation is guaranteed to have computed the correct answer if it generates a trajectory that satisfies equation (1-5) and has matched endpoints.   This is true even if the simulation has been driven by an intelligent goal oriented algorithm rather than a conventional random number generator. Shaped simulation is based on the idea of replacing conventional random number generators by appropriate goal oriented algorithms to improve both speed and accuracy.

Another set of open questions concerns the class of formal algebraic objects known as t-loops.   These are introduced in Section 4.7.4 to facilitate the discussion of matched endpoints for discrete time trajectories.   Section 7.5 extends the basic concept from discrete to continuous time. T-loops appear to fall within the domain of renewal theory. They may also possess other properties that make them interesting formal objects in their own right.

Other areas of interest include extending observational stochastics to deal with trans-distributional quantities of the type described in Section 7.9.3, formalizing sensitivity analyses (Suri 1983) of the type introduced in Section 1.4.6, and transforming the distributional parameters and assumptions commonly employed in traditional stochastic models into the observational parameters used in Chapters 5, 6 and 7. Extending observational stochastics to deal with transient behavior may also be possible by applying the proportionalist interpretation of chance developed in Chapters 8 and 9. Many other classical results from the theory of probability are likely to have simple observational counterparts that can also be derived through proportionalist arguments.

1.6 Organizational notes

Textbooks that deal with mathematical topics are typically organized around sets of powerful mathematical theorems. Traditionally the statement of each theorem is followed by a formal proof, which is then followed by case studies that illustrate the theorem's application to various problems.

This book follows a different path. Most chapters begin with a series of examples, starting with very simple cases and then adding layers of complexity to illustrate the broadly applicable nature of the original analysis. This streamlines the notation required at each stage and focuses the discussion on readily understandable insights rather than powerful but abstract mathematical theorems. A few generalized theorems are also presented, but only after simpler specific cases have been explained thoroughly.

Readers with advanced mathematical backgrounds may find this example-oriented format a bit too leisurely. These readers may wish to proceed directly to Section 3.7 for a rigorous characterization of observational stochastics.   Two powerful theorems and two important corollaries that can be derived within this formal framework are presented in Chapters 4 and 7:

Theorem 4.1 in Section 4.6 presents an important relationship that is valid for all discrete time trajectories with matched endpoints.

Corollary 4.1 extends Theorem 4.1 to the completely general case of discrete time trajectories with unmatched endpoints.

Theorem 7.1 in Section 7.3 presents an important relationship valid for all continuous time trajectories with matched endpoints.

Corollary 7.1 extends Theorem 7.1 to the completely general case of continuous time trajectories with unmatched endpoints.

These theorems and corollaries, along with the formal LCD model presented in Section 3.7, represent the essential mathematical core of this book.

A number of other chapters present new derivations of classical results that are well known within traditional stochastic contexts. In particular, several standard results from queuing theory including Little's Law, the steady state distribution of the M/M/1 queue, and various results for cyclic queues are presented in Sections 7.8 and 7.9.   In addition, an intuitive, example-based treatment of the Law of Large Numbers is presented in Section 9.1.

The motivation for the Ergodic Theorem is also examined through a series of intuitive examples in Section 9.2.   These examples illustrate the manner in which observational stochastics can be extended to deal with transient aspects of system behavior.

CHAPTER 2
Traditional Random Walks

2.1 Overall goals of this chapter

Chapter 1 introduced the concept of a simple random walk and discussed several implications of equations (1-1) through (1-4). However, these equations were not derived in a rigorously formal mathematical sense. This chapter presents a derivation within the context of traditional stochastic analysis. A number of basic concepts from the theory of probability and stochastic processes are reviewed in greater detail than is customary. Readers already familiar with these aspects of stochastic modeling may wish to skip this chapter entirely and proceed to Chapter 3.

2.2 Deterministic state transition diagrams

To begin the analysis, note that the transitions that take place during the random walk discussed in Chapter 1 can be represented by the state transition diagram shown in Figure 2-1. The circles in this figure represent the four stations, and the arrows represent the possible station-to-station transitions that can take place during a single step. Even though the walk is classified as random, the diagram in Figure 2-1 is clearly deterministic: if the walker is currently at station 1, and if the next turn is to the right, the walker will always move to station 2.

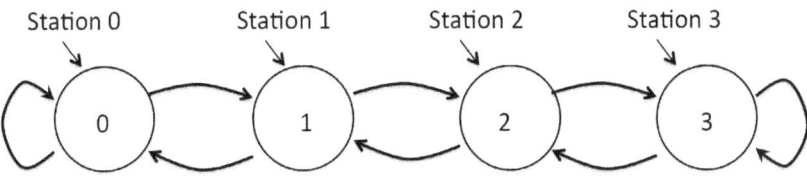

Figure 2-1. Deterministic State Transition Diagram

All the other transitions shown in Figure 2-1 are equally deterministic. Most "random" systems that are analyzed using stochastic models have a deterministic core that can be specified using a state transition diagram of the type illustrated in Figure 2-1.

2.3 The traditional characterization of uncertainty

The next issue to consider is the nature of the mechanism that causes the walker to move from station to station as the walk proceeds.   In order for the walk to be random, the sequence of left and right turns that directs the walker's movements from station to station must be inherently uncertain.   One way to represent this type of uncertainty is to assume that the walker tosses a coin before leaving each station. If the coin comes up heads, the next turn is to the right; if it comes up tails, the next turn is to the left.

2.3.1 Sample spaces and probabilities

The abstract notion of sets and their members provides the standard mathematical foundation for dealing with uncertainty. Essentially, the members of a set are used to represent the possible outcomes of a process that is inherently uncertain.   Such sets of possible outcomes are referred to as *sample spaces*.

In the case of a random walk, the set {heads, tails} is used to represent the possible outcomes of each coin toss.   There is, of course, a direct connection between the observable outcomes of an actual coin toss and the members of this sample space. However, the set {heads, tails} can also be regarded as an abstract entity whose existence is entirely independent of the physical world.   As such, its only properties are those defined within the formal domain of set theory: union, intersection, and so on.

The next step in the conventional representation of uncertainty is to associate a probability with each member and each subset (within a sample space. A probability is simply a non-negative real number. The probability associated with an entire sample space must be equal to 1. Also, the probabilities that are assigned to subsets must satisfy the simple and entirely straightforward rules of measure theory: most importantly, if two subsets are disjoint, the probability assigned to their union must be equal to the sum of their individual probabilities.

Technically speaking, there is no need to specify a connection between probabilities and observable properties of processes that operate in the real world. This connection lies beyond the formal boundaries of the theory of probability. However, as discussed in Chapter 8, Bayesians and frequentists have clear but differing views regarding the connection between abstract mathematical probabilities and observable properties of processes that operate in the real world. These views are based on principles that have strong intuitive appeal, but neither view can be proven correct in any rigorous sense. Simply put, the theory of probability as formalized by Komogorov (1933) imposes no assumptions about the connection between probabilities and real world observables.

2.3.2 Random variables

A sample space, together with its probability distribution, is referred to as a random variable. Most of the results derived from probabilistic models are expressed in terms of random variables and their associated probability distributions.

For the random walk being discussed here, the outcome of each successive coin toss is represented by a separate random variable. For each of these random variables, the sample space is the set {heads, tails}, the probability associated with heads is equal to r, and the probability associated with tails is equal to 1-r. This sequence of identically distributed random variables, together with the relationships represented in Figure 2-1, provides the basis for a traditional stochastic model of a random walk.

2.4 Equations for the state of a random walk

Following the description in Chapter 1, suppose the walk begins with the walker at station 2. The first turn will take the walker to station 3 or station 1, depending on the outcome of the first coin toss. Since the outcome of the coin toss is uncertain, it makes sense to represent the walker's station after the first turn, and after each successive turn, by a sequence of random variables. Specifically, assume that S(1) represents the walker's station after the first turn, S(2) represents the walker's station after the second turn, and so on.

Each random variable S(n) has the same sample space: {0, 1, 2, 3}. To complete the specification of S(n), it is necessary to assign a probability to 0, 1, 2 and 3 for each value of n. These probabilities will be denoted by $\underline{P}(0,n)$, $\underline{P}(1,n)$, $\underline{P}(2,n)$, and $\underline{P}(3,n)$. Formally speaking, $\underline{P}(k,n)$ is the probability that the random variable S(n) is equal to k. Note that $\underline{P}$ is underlined to emphasize that $\underline{P}(k,n)$ represents a probability rather than a directly observable proportion.

2.4.1 Position of the walker after the first coin toss

Assuming the walk begins at station 2, the probability of the walker being at station 3 after the first turn is equal to r, and the probability of the walker being at station 1 after the first turn is equal to 1-r. The probability of being at stations 0 or 2 is zero. The values of $\underline{P}(k,1)$ must thus be assigned as follows:

$$\underline{P}(0,1) = 0 \tag{2-1}$$

$$\underline{P}(1,1) = 1 - r \tag{2-2}$$

$$\underline{P}(2,1) = 0 \tag{2-3}$$

$$\underline{P}(3,1) = r \tag{2-4}$$

2.4.2 Position of the walker after the second coin toss

Now consider the random variable that represents the position of the walker after the second turn. Note that $\underline{P}(0,2)$ represents the probability that the walker is at station 0 after the second turn. As shown in Figure 2-1, there are two ways this event can occur: either the walker is at station 1 after the first turn and the outcome of the second coin toss is tails (causing the walker to turn left), or the walker is at station 0 after the first turn and the outcome of the second coin toss is tails (causing the walker to turn left, rebound off the barrier, and return to station 0). However, $\underline{P}(0,1)$ is equal to zero by equation (2-1), so only the first option is physically possible.

It is reasonable to assume that the random variable representing the outcome of each coin toss and the random variable representing the walker's current location are statistically independent. This assumption is implicit in the traditional stochastic model of a random walk. In this case, the assumption of statistical independence implies that $\underline{P}(0,2)$ is given as follows:

$$\underline{P}(0,2) = \underline{P}(1,1) \times (1-r)$$

$$= (1-r) \times (1-r)$$

$$= 1 - 2r + r^2 \qquad (2\text{-}5)$$

Now consider $\underline{P}(1,2)$, the probability that the walker is at station 1 after the second turn. There are two ways this event can occur: either the walker turns right when exiting from station 0, or left when exiting from station 2. However, the probabilities of the walker being in stations 0 or 2 after the first turn are both zero, as indicated by equations (2-1) and (2-3). Thus

$$\underline{P}(1,2) = 0 \qquad (2\text{-}6)$$

The next step is to evaluate $\underline{P}(2,2)$. Note that station 2 can be reached from station 1 by turning right, or from station 3 by turning left. Assuming once again that the random variables representing the walker's location and the outcome of the coin toss are statistically independent, $\underline{P}(2,2)$ must have the following form:

$$\underline{P}(2,2) = \underline{P}(1,1) \times r + \underline{P}(3,1) \times (1-r)$$

$$= (1-r) \times r + r \times (1-r)$$

$$= 2r - 2r^2 \qquad\qquad (2\text{-}7)$$

The last probability to consider is $\underline{P}(3,2)$. By equation (2-3), $\underline{P}(2,1)$ is equal to 0. Thus the only way the walker can arrive at station 3 after the second turn is by starting from station 3 and turning right to rebound off the reflecting barrier. This implies

$$\underline{P}(3,2) = \underline{P}(3,1) \times r$$

$$= r \times r$$

$$= r^2 \qquad\qquad (2\text{-}8)$$

Note that $\underline{P}(0,2) + \underline{P}(1,2) + \underline{P}(2,2) + \underline{P}(3,2) = 1$ for all values of r.

2.4.3 Position of the walker after the nth coin toss

The arguments used to derive equations (2-5) through (2-8) can be generalized to express the position of the walker after n turns as a function of the walker's position after $n-1$ turns.

Let $\underline{P}(0,n)$ represent the probability that the walker is located at station 0 after n turns. To derive an expression for $\underline{P}(0,n)$, consider all ways the walker could possibly reach station 0 in a single step. One option is to start at station 0 and turn left, which causes the walker to rebound off the reflecting barrier. The probability the walker reaches station n by this route is $\underline{P}(0,n-1)\times(1-r)$. The other option is to start at station 1 and turn left, which of course has an associated probability of $\underline{P}(1,n-1)\times(1-r)$.

These are the only two options. They represent separate (disjoint) events, which implies that the probability either will occur is simply the sum of the two associated probabilities. Thus

$$\underline{P}(0,n) = \underline{P}(0,n-1)\times(1-r) + \underline{P}(1,n-1)\times(1-r) \tag{2-9}$$

Similar arguments can be applied to stations 1, 2 and 3. These lead to the following equations:

$$\underline{P}(1,n) = \underline{P}(0,n-1)\times r + \underline{P}(2,n-1)\times(1-r) \tag{2-10}$$

$$\underline{P}(2,n) = \underline{P}(1,n-1)\times r + \underline{P}(3,n-1)\times(1-r) \tag{2-11}$$

$$\underline{P}(3,n) = \underline{P}(2,n-1)\times r + \underline{P}(3,n-1)\times r \tag{2-12}$$

Equations (2-9) through (2-12) can be applied recursively to generate a sequence of random variables that characterize the walker's position after each turn.

To complete the specification of this model, it is necessary to identify the location of the walker at the start of the walk. In this particular case, assume the walker begins at station 2. Zero turns have been made at this point. Thus, the walker's initial position is characterized by the random variable $\underline{P}(j,0)$ whose distribution is given as follows:

$$\underline{P}(0,0) = 0 \tag{2-13}$$

$$\underline{P}(1,0) = 0 \tag{2-14}$$

$$\underline{P}(2,0) = 1 \tag{2-15}$$

$$\underline{P}(3,0) = 0 \tag{2-16}$$

Even though the walker's initial position is characterized by a random variable, equation (2-15) implies that the walker is always at station 2 when the walk begins. There is no uncertainty about this fact.

2.4.4 The underlying stochastic process

Equations (2-9) through (2-12), together with the initial distribution specified by equations (2-13) through (2-16), determine the probability distribution of each random variable in the sequence S(0), S(1), S(2) ... . Sequences of this type are referred to as stochastic processes.

In general, stochastic processes are specified as the solution to sets of equations such as equations (2-9) through (2-16). In the case of a random walk, the state transition diagram in Figure 2-1 and the assumption that the outcome of each coin toss is statistically independent of the walker's current state provide useful insights into these equations; however, they are not necessary in any formal sense. The mathematical relationships that are expressed by equations (2-9) through (2-16) are sufficient to characterize this particular stochastic process. Nothing further is required.

To explore the limiting behavior of this stochastic process as n increases, it is helpful to begin by assigning a specific numerical value to r. Suppose that the value of r in equations (2-9) through (2-12) is set equal to 2/3. The probability distributions for the random variables that make up this stochastic process can then be generated by a routine series of iterations. The first fifteen probability distributions generated by these iterations are presented in Table 2-1.

n	P(0,n)	P(1,n)	P(2,n)	P(3,n)
0	.000	.000	1.000	.000
1	.000	.333	.000	.667
2	.111	.000	.444	.444
3	.037	.222	.148	.593
4	.086	.074	.346	.494
5	.053	.173	.214	.560
6	.075	.107	.302	.516
7	.061	.151	.243	.545
8	.071	.122	.282	.526
9	.064	.141	.256	.539
10	.068	.128	.274	.530
11	.066	.137	.262	.536
12	.067	.131	.270	.532
13	.066	.135	.265	.534
14	.067	.132	.268	.533
15	.066	.134	.266	.534

Table 2-1. Values of $\underline{P}(k,n)$ for $n = 0, 1, \ldots 15$

The probability distributions displayed in the first few rows of this table vary considerably. However, the distributions (rounded to three significant digits) begin to stabilize towards the bottom of the table. This is easily seen by scanning down each column. Within each column, the values can be seen moving steadily closer to the limiting values displayed in Table 2-2 (and derived in Section 2.5). If Table 2-1 were extended further, all additional rows would contain entries that are within .001 of these limiting values.

P(0,n) converges to 1/15 = .0666666...	as n approaches infinity
P(1,n) converges to 2/15 = .1333333...	as n approaches infinity
P(2,n) converges to 4/15 = .2666666...	as n approaches infinity
P(3,n) converges to 8/15 = .5333333...	as n approaches infinity

Table 2-2. Limiting values of P(k,n)

2.5 The steady state distribution

As long as n remains finite, the limiting values shown in Table 2-2 are never attained exactly. Nevertheless, if n is at least moderately large, the values of P(j,n) will all be quite close to their respective limits. For large values of n, the differences are so small as to be inconsequential.

These comments make it interesting to analyze a different stochastic process that is closely related to the one considered in Table 2-1.   Equations (2-9) through (2-12) once again define the step-by-step dynamics of this new process, and the value of r is still equal to 2/3. However, the initial values of P(0,0), P(1,0), P(2,0) and P(3,0) are no longer specified by equations (2-13) through (2-16).   Instead, as shown in Table 2-3, the initial probability distribution for this new stochastic process is set equal to the limiting distribution specified in Table 2-2.

This change of initial conditions has important consequences. As shown in Table 2-4, every row in the recomputed version of Table 2-1 is now exactly equal to the row above it. In other words, the random variables that make up this related stochastic process are identically distributed.   Stochastic processes with this special property are said to be *stationary* or in *steady state*.   The common probability distribution shared by all the random variables associated with such a stochastic process is referred to as a stationary or steady state distribution.

$\underline{P}(0,0) = 1/15 = .066...$
$\underline{P}(1,0) = 2/15 = .133...$
$\underline{P}(2,0) = 4/15 = .266...$
$\underline{P}(3,0) = 8/15 = .533...$

Table 2-3. New Initial Conditions

Not all stochastic processes converge to a limiting distribution of the type illustrated in Table 2-4.

n	$\underline{P}(0,n)$	$\underline{P}(1,n)$	$\underline{P}(2,n)$	$\underline{P}(3,n)$
0	.066...	.133...	.266...	.533...
1	.066...	.133...	.266...	.533...
2	.066...	.133...	.266...	.533...
3	.066...	.133...	.266...	.533...

Table 2-4. Values of $\underline{P}(k,n)$ under New Initial Conditions

Even when convergence does occur, there is no guarantee that all possible initial distributions will converge to the same limiting distribution. Stochastic processes that always converge to the same limiting distribution, regardless of initial conditions, are classified as *ergodic*.

The stochastic process defined by equations (2-9) through (2-16) is, in fact, ergodic. When r is set equal to 2/3, ergodicity implies that the initial probability distribution specified in line 1 of Table 2-1 is unimportant. Regardless of the starting point, the distributions in Table 2-1 always converge to the same limiting values as the number of lines in the table grows large.

The next task is to generalize this discussion from the case where r is equal to 2/3 to the case where r remains a symbolic parameter. That is, the goal is to express the steady state distribution as an algebraic function of the symbolic variable r.

Suppose the desired steady state distribution is represented symbolically as: $\underline{P}(0)$, $\underline{P}(1)$, $\underline{P}(2)$, $\underline{P}(3)$. As before, equations (2-9) through (2-12) specify how the probability distribution after n steps is related to the probability distribution after n-1 steps. However, when the stochastic process is in steady state, both these distributions will be identical: $\underline{P}(0,n)$ and $\underline{P}(0,n-1)$ will both be equal to $\underline{P}(0)$; $\underline{P}(1,n)$) and $\underline{P}(1,n-1)$ will both be equal to $\underline{P}(1)$; and so on.

Under these conditions, equations (2-9) through (2-12) are transformed immediately into equations (2-17) through (2-20).

$$\underline{P}(0) = \underline{P}(0) \times (1-r) + \underline{P}(1) \times (1-r) \tag{2-17}$$

$$\underline{P}(1) = \underline{P}(0) \times r + \underline{P}(2) \times (1-r) \tag{2-18}$$

$$\underline{P}(2) = \underline{P}(1) \times r + \underline{P}(3) \times (1-r) \tag{2-19}$$

$$\underline{P}(3) = \underline{P}(2) \times r + \underline{P}(3) \times r \tag{2-20}$$

Explicit expressions for $\underline{P}(0)$, $\underline{P}(1)$, $\underline{P}(2)$ and $\underline{P}(3)$ can be derived from equations (2-17) through (2-20). Note that equation (2-17) implies that $\underline{P}(1)$ can be expressed as a simple function of $\underline{P}(0)$.

$$P(1) = \underline{P}(0) \times r / (1-r) \tag{2-21}$$

Combining equations (2-18) and (2-21) makes it possible to express $\underline{P}(2)$ as a simple function of $\underline{P}(0)$.

$$P(2) = \underline{P}(0) \times [r / (1-r)]^2 \tag{2-22}$$

Similarly, equations (2-19), (2-21) and (2-22) make it possible to express $\underline{P}(3)$ as a simple function of $\underline{P}(0)$.

$$P(3) = \underline{P}(0) \times [r / (1-r)]^3 \tag{2-23}$$

To complete this analysis, it is necessary to obtain an expression for $\underline{P}(0)$. Note first that the sum $\underline{P}(0) + \underline{P}(1) + \underline{P}(2) + \underline{P}(3)$ must be equal to 1. All probability distributions must satisfy a normalization constraint of this type. This particular normalization constraint, along with equations (2-21) through (2-23), yields:

$$\underline{P}(0) = \frac{1}{1 + r/(1-r) + [r/(1-r)]^2 + [r/(1-r)]^3} \tag{2-24}$$

When r = 2/3, set of the numerical values obtained from equations (2-21) through (2-24) are identical to the steady state distribution that appears in each row of Table 2-4.

2.6 Implications and applications

Equations (2-21) through (2-24) have exactly the same algebraic structure as the solution given by equations (1-1) through (1-4) in Chapter 1. There are, however, some differences. The most obvious difference is that the letter P is not underlined in Chapter 1, but it is here. In addition, the symbolic variable R that appears in Chapter 1 is replaced by r. These cosmetic differences are only the tip of the iceberg. The actual differences between the two sets of equations are fundamental and profound.

Note first that the mathematically rigorous analysis used to derive equations (2-21) through (2-24) is only loosely connected to the original problem stated in Chapter 1.   The original problem deals with a random walk that is assumed to take place in the real world. This random walk is associated with a trajectory of the type illustrated in Figure 1-4.   It is assumed that certain properties of this walk are directly observable: the overall proportion of right turns, which is represented symbolically by R, and the proportion of visits the walker makes to each of the four stations, which are represented symbolically by P(0), P(1), P(2) and P(3).   The original problem is to express the values of P(j) as functions of R.

In this chapter, the random walk has been modeled by a stochastic process.   The structure of this abstract mathematical entity is formally specified by equations (2-9) through (2-16), and its steady state distribution is given by equations (2-21) through (2-24).

At this point it is entirely appropriate for a practitioner to question what connection, if any, exists between the solution expressed by equations (2-21) through (2-24) and the original problem stated in Chapter 1.   These equations characterize the steady state distribution of a mathematical abstraction known formally as an ergodic stochastic process. The symbolic variable r is a parameter of this stochastic process, and $\underline{P}(0)$, $\underline{P}(1)$, $\underline{P}(2)$, and $\underline{P}(3)$ are its stationary (i.e., steady state) distribution. The relationship between the original problem and this mathematical solution is not immediately apparent.

2.7 The Law of Large Numbers and the Ergodic Theorem

The symbols r and $\underline{P}(j,n)$ that appear in equations (2-21) through (2-24) represent probabilities.   As mentioned in Section 2.3.1, probabilities are mathematical abstractions: real numbers that lie in the interval [0,1] and are associated with members of a sample space.   Their formal definitions must conform to the axioms of measure theory, but these definitions are not connected directly to the observable properties of any process or system that operates in the real world.

The challenge for practitioners is to link these formal mathematical abstractions to real world entities that can be observed and measured directly. Such linkages depend, in general, on assumptions regarding the underlying factors that control the behavior of real world systems. For the simple random walk being considered here, the abstract probabilities r and $\underline{P}(j,n)$ can be linked to the observable proportions R and P(j) by introducing assumptions regarding the mechanism that determines the direction of each turn the walker makes.

It is convenient to characterize such mechanisms in terms of idealized random number generators. Modern random number generators use unstable digital circuits and special algorithms to produce sequences of real numbers that are uniformly distributed over the interval [0,1]. The detailed properties of these circuits, algorithms and sequences are not important for this discussion. The only requirement is the assumption that such idealized random number generators can, in principle, exist.

Given the existence of such a random number generator, the parameter r that appears in equations (2-21) through (2-24) can be easily linked to the direction the walker actually turns: if the next value produced by the idealized random number generator is less than or equal to r, the next turn is to the right. If not, the next turn is to the left.

This linkage makes it possible to examine the relationship between the probability r that appears in equations (2-21) through (2-24) and the observable variable R that appears in equations (1-1) through (1-4). Not surprisingly, R is likely to be close to r. This likelihood improves steadily as the length of the walk increases. In the limit, the probability that R and r differ by more than a negligible amount approaches zero. This conclusion follows from an immensely important classical result from probability theory: the Law of Large Numbers. Chapter 9 presents a formal derivation and detailed discussion of this result.

Under this same set of assumptions, it is also true that each observed proportion P(j) is likely to converge to the corresponding steady state probability $\underline{P}$(j) as the length of the random walk approaches infinity. The Law of Large Numbers is not applicable in this case because the sequence of random variables that comprise a stationary stochastic process need not be statistically independent. A more powerful and nuanced result – the Ergodic Theorem – is required to prove that the observed proportion P(j) will converge *almost surely* to the associated steady state probability $\underline{P}$(j) in such cases.

The Ergodic Theorem is a fundamentally important result.   It makes good sense on an intuitive level and provides the bridge that connects the formal mathematical theory of stochastic processes with the observable properties of systems and processes that operate in the real world.   Chapter 9 presents a practitioner-oriented introduction to this theorem and related concepts including Borel measure, sets of measure zero, and – for interested readers – a celebrated example of mathematical exotica: the Cantor ternary set, which is a set of measure zero containing subsets that are not Borel measurable.

Most practitioners can safely ignore the esoteric mathematical concepts explored in Chapter 9. However, all practitioners should be aware of the main implications of the Law of Large Numbers and the Ergodic Theorem. These are:

1. The longer the trajectory, the greater the likelihood that equations (2-9) through (2-13) will yield accurate predictions for

the observed values of P(j).

2.   Even if the length of the trajectory is allowed to approach infinity, the accuracy of these predictions can never be guaranteed with absolute certainty.

3. Points 1 and 2 can only be justified in cases that are intrinsically random: in this example, cases where there are good reasons to believe that the direction of each successive turn can be associated with a sequence of values returned by an idealized random number generator.   Certain physical processes such as coin tossing satisfy this requirement. On the other hand, the assumption of intrinsic randomness is difficult to justify when the direction of each turn is controlled by the walker's opinion of songs on a playlist (as discussed in Section 1.4.5).

More often than not, it has been this author's experience that the assumption of intrinsic randomness requires a substantial leap of faith – especially when the systems being modeled are driven ultimately by human actions and decisions (e.g., communication networks, internet/web servers, highway traffic, purchasing/voting decisions,   etc.) Appeals to the Law of Large Numbers and the Ergodic Theorem are difficult to justify in such cases and represent a legitimate risk for practitioners.

Despite these concerns, results derived from ergodic stochastic models can be remarkably accurate in practice.   The analysis presented in Chapter 3 explains why this is often so.

CHAPTER 3
LCD Model of a Random Walk

3.1 Characterization of observability

As discussed in Chapter 1, the variables used in observational stochastics represent quantities that are directly observable. This simple concept is easy to understand on an intuitive level. However, characterizing observability in terms that are both precise and broadly applicable is a challenging task. This is one of the main objectives of Chapter 3.

The analysis in the first part of this chapter deals with the simple random walk discussed in Chapters 1 and 2. A definition of observability within this context is developed in Section 3.2. This definition reflects the standard procedures that practitioners routinely employ when measuring actual systems. Sections 3.3 through 3.6 build upon this definition by presenting a formal analysis of random walks based entirely on the principles of observational stochastics.

Simple random walks also serve as a springboard for the general treatment of observability developed in Section 3.7. The discussion in Section 3.7 begins with the specification of an entity known as a *loosely constrained deterministic (LCD) model*. These LCD models (Buzen 2012) are used to develop a precise and broadly applicable characterization of observability. Observational stochastics rest upon the formal foundation that these models provide.

3.2 Observable variables for random walks

The state transition diagram in Figure 2-1 can be regarded as representing the rules that govern a random walk. These rules specify the next station the walker will encounter after making a left or right turn. Note that the state transition diagram can also be used to infer information about the direction of each turn when only a trajectory such as the one shown in Figure 1-4 is available. For example, if station 2 is followed immediately by station 3 in a trajectory, the analyst can infer that the walker turned to the right after this particular departure from station 2.

The notion of turning to the right has meaning in the real world. On the other hand, the fact that station 2 is followed immediately by station 3 is a property of the trajectory itself. A trajectory for this random walk is simply a mathematical function that maps a sequence of integers – representing the step-by-step passage of time – into a set of states comprised of the integers 0 through 3.

In this specific case, suppose an analyst is presented with a trajectory of the type shown in Figure 1-4. Assume the analyst knows that this trajectory was, in fact, generated by a random walk that conforms to the rules represented by the state transition diagram in Figure 2-1. It is then a routine matter for the analyst to extract the following values from the trajectory:

$R(j)^{\#}$ = number of times the walker leaves station j and turns right
$\quad (j = 0,1,2,3)$

$L(j)^{\#}$ = number of times the walker leaves station j and turns left
$\quad (j = 0,1,2,3)$

The hash tags incorporated into the symbols $R(j)^{\#}$ and $L(j)^{\#}$ highlight the fact that these variables represent basic measurements (i.e., raw counts) that can be regarded as directly observable properties of a trajectory.

Other observable quantities can now be expressed as simple functions of $R(j)^{\#}$ and $L(j)^{\#}$.

R(j) = proportion of time the walker turns to the right when exiting from station j

$$= \frac{R(j)^{\#}}{R(j)^{\#} + L(j)^{\#}} \qquad (3\text{-}1)$$

V(j) = total number of visits the walker makes to station j

$$= R(j)^{\#} + L(j)^{\#} \qquad (3\text{-}2)$$

Note that this definition of V(j) excludes the final visit the walker makes to the last station in the trajectory (since the walker never leaves the final station).   Excluding this final visit simplifies the form of the mathematical solution that is ultimately derived. Chapter 4 presents a general procedure for adjusting the solution derived here to account for this final visit.   The definitions of V, R and P(j) that are presented below also exclude the final visit.

V   = total number of visits the walker makes to all stations (excluding the final visit)

  = total number of turns the walker makes during the entire walk

$$= V(0) + V(1) + V(2) + V(3) \qquad (3\text{-}3)$$

R   = proportion of time the walker turns to the right during the ` entire walk

$$= \frac{R(0)^{\#} + R(1)^{\#} + R(2)^{\#} + R(3)^{\#}}{V} \qquad (3\text{-}4)$$

P(j) = proportion of visits to station j during the entire walk

$$= \frac{V(j)}{V} \qquad (3\text{-}5)$$

3.3 Balance equations

The objective of this analysis is to express the values of P(j) as functions of R.   The derivation depends on an important assumption that has not been considered thus far. Every trajectory that is analyzed within the framework of observational stochastics is assumed to correspond to a sequence of observations that are made over a single contiguous interval of time.   There must be no gaps or discontinuities. If discontinuities exist, each contiguous sub-interval must be analyzed separately.

This requirement has important implications. Imagine following the walker from station to station over the course of the walk. As the walk proceeds, the walker enters a station, spends an unspecified amount of time there, and then exits. This implies that the total number of entrances into each station must be exactly equal to the total number of exits from that station.   There are only two minor exceptions to this general rule: there is one extra exit from the initial station at the start of the walk, and one extra entrance into the final station at the end of the walk. These simple and intuitively obvious relationships are valid for any trajectory that represents a single contiguous interval of time.

If the initial and final stations are the same, the extra exit from the initial station is balanced by the extra entrance into the final station.   When the two endpoints are matched in this manner, the number of entrances must be exactly equal to the number of exits for the initial and final stations as well.

Consider the implications of matched endpoints for station 0. As illustrated in Figure 2-1, there are two ways the walker can enter station 0: by exiting from station 0 and turning left (rebounding off the reflecting barrier), or by exiting from station 1 and turning left. The observable variables $L(0)^{\#}$ and $L(1)^{\#}$ represent the number of times these two events occur over the course of the trajectory.

Now consider the number of times the walker exits from station zero. Upon exiting, the walker can turn either right or left. The observable variables $R(0)^{\#}$ and $L(0)^{\#}$ represent the number of times these events occur over the course of the trajectory. If the endpoints are matched, the number of exits must be equal to the number of entrances for all stations, including station 0. Thus

$$R(0)^{\#} + L(0)^{\#} = L(0)^{\#} + L(1)^{\#} \qquad (3\text{-}6)$$

Similar arguments can also be applied to stations 1, 2 and 3. These arguments yield the following equations:

$$R(1)^{\#} + L(1)^{\#} = R(0)^{\#} + L(2)^{\#} \qquad (3\text{-}7)$$

$$R(2)^{\#} + L(2)^{\#} = R(1)^{\#} + L(3)^{\#} \qquad (3\text{-}8)$$

$$R(3)^{\#} + L(3)^{\#} = R(2)^{\#} + R(3)^{\#} \qquad (3\text{-}9)$$

Equation (3-2) makes it possible to modify equations (3-6) through (3-9) as follows.

First replace the term $R(j)^{\#} + L(j)^{\#}$ by the term $V(j)$ on the left hand sides of these equations. Then replace all occurrences of the term $L(j)^{\#}$ by the term $[V(j) - R(j)^{\#}]$. These substitutions result in equations (3-10) through (3-13).

$$V(0) = [V(0) - R(0)^{\#}] + [V(1) - R(1)^{\#}] \qquad (3\text{-}10)$$

$$V(1) = R(0)^{\#} + [V(2) - R(2)^{\#}] \qquad (3\text{-}11)$$

$$V(2) = R(1)^{\#} + [V(3) - R(3)^{\#}] \qquad (3\text{-}12)$$

$$V(3) = R(2)^{\#} + R(3)^{\#} \qquad (3\text{-}13)$$

Equations (3-1), (3-2) and (3-5) can now be used to further simplify equations (3-10) through (3-13).   Replace V(j) by $P(j) \times V$ for each value of j (0, 1, 2 and 3). Then replace $R(j)^{\#}$ by $R(j) \times P(j) \times V$ for each value of j (0, 1, 2 and 3).   These substitutions yield:

$$P(0) \times V = [P(0) \times V - R(0) \times P(0) \times V] + [P(1) \times V - R(1) \times P(1) \times V]$$
$$(3\text{-}14)$$

$$P(1) \times V = R(0) \times P(0) \times V + [P(2) \times V - R(2) \times P(2) \times V] \qquad (3\text{-}15)$$

$$P(2) \times V = R(1) \times P(1) \times V + [P(3) \times V - R(3) \times P(3) \times V] \qquad (3\text{-}16)$$

$$P(3) \times V = R(2) \times P(2) \times V + R(3) \times P(3) \times V \qquad (3\text{-}17)$$

Dividing both sides of each equation by V and factoring out common terms yields:

$$P(0) = P(0) \times [1 - R(0)] + P(1) \times [1 - R(1)] \qquad (3\text{-}18)$$

$$P(1) = P(0) \times R(0) + P(2) \times [1 - R(2)] \qquad (3\text{-}19)$$

$$P(2) = P(1) \times R(1) + P(3) \times [1 - R(3)] \qquad (3\text{-}20)$$

$$P(3) = P(2) \times R(2) + P(3) \times R(3) \qquad (3\text{-}21)$$

Equations (3-18) through (3-21) will be referred to as the balance equations for this random walk.   They are valid for any trajectory with matched endpoints (i.e., whose initial and final stations are the same). No other assumptions are required.

3.4 Solution with empirical independence

Assume next that the trajectory being considered also satisfies the assumption of empirical independence specified by equation (1-5) in Chapter 1. In particular assume that $R(0) = R(1) = R(2) = R(3)$.

It follows immediately that all four values of R(j) must be equal to R. Replacing all occurrences of R(j) by R in equations (3-18) through (3-21) yields:

$$P(0) = P(0) \times (1-R) + P(1) \times (1-R) \tag{3-22}$$

$$P(1) = P(0) \times R + P(2) \times (1-R) \tag{3-23}$$

$$P(2) = P(1) \times R + P(3) \times (1-R) \tag{3-24}$$

$$P(3) = P(2) \times R + P(3) \times R \tag{3-25}$$

These simplified balance equations, which reflect the impact of empirical independence, have exactly the same structure as equations (2-17) through (2-20) in Chapter 2. The steps that led to the solution given by equations (2-21) through (2-24) can be replicated in this case to yield:

$$P(0) = \frac{1}{1 + [R/(1-R)] + [R/(1-R)]^2 + [R/(1-R)]^3} \tag{3-26}$$

$$P(1) = P(0) \times R/(1-R) \tag{3-27}$$

$$P(2) = P(0) \times [R/(1-R)]^2 \tag{3-28}$$

$$P(3) = P(0) \times [R/(1-R)]^3 \tag{3-29}$$

Even though equations (3-26) through (3-29) have the same form as equations (2-21) through (2-24), their interpretation is clearly different. The variables R and P(j) that appear in equations (3-26) through (3-29) do not represent probabilities. Instead, they represent observable quantities whose definitions are specified by equations (3-1) through (3-5). Note that these definitions are expressed in terms of $R(j)^{\#}$ and $L(j)^{\#}$. These values can be extracted directly from trajectories such as the one illustrated in

Figure 1-4.   Thus, equations (3-26) through (3-29) can be applied directly to such trajectories without appealing to the Law of Large Numbers or the Ergodic Theorem.

Note also that equations (3-26) through (3-29) are guaranteed to generate values of P(j) that agree exactly with observed values for any trajectory with matched endpoints that satisfies the assumption of empirical independence.   A generalization of these equations to cases where trajectories do not satisfy empirical independence, but still have matched endpoints, is presented in the next section. Chapter 4 describes a method for generalizing these results further by removing the requirement that endpoints be matched.

3.5 Solution without empirical independence

For the class of trajectories with matched endpoints that satisfy the assumptions of empirical independence, the original balance equations [(3-18) through (3-21)] simplify immediately and can be rewritten as equations (3-22) through (3-25). As shown in Section 3.4, the simplified balance equations can then be solved to obtain expressions for the values of P(j) as functions of R.   These expressions are given by equations (3-26) through (3-29).

Now consider a more general class of trajectories where endpoints are still matched, but where the assumption of empirical independence is not necessarily satisfied.   The original balance equations (3-18) through (3-21) are, of course, valid for such trajectories.   Moreover, this set of slightly more complex linear equations can again be solved to yield expressions for the values of P(j). To express the solution in a compact form, first define the auxiliary variables Q(0), Q(1) and Q(2) as follows:

$$Q(n) = R(n)/[1 - R(n+1)] \quad \text{for } n = 0, 1 \text{ and } 2$$

The values of P(j) for j = 0, 1, 2 and 3 can then be expressed as follows.

$$P(0) = \frac{1}{1+Q(0)+Q(0)\times Q(1)+Q(0)\times Q(1)\times Q(2)} \qquad (3\text{-}30)$$

$$P(1) = P(0)\times Q(0) \qquad (3\text{-}31)$$

$$P(2) = P(0)\times Q(0)\times Q(1) \qquad (3\text{-}32)$$

$$P(3) = P(0)\times Q(0)\times Q(1)\times Q(2) \qquad (3\text{-}33)$$

The derivation of these equations corresponds directly to the step-by-step procedure used in the simpler case where all values of R(j) are equal to R. The only difference is that it is now necessary to keep track of each individual value of R(j). Thus, the following changes must be made to equations (3-26) through (3-29):

$R/(1\text{-}R)$ must be replaced by $Q(0)$

$[R/(1\text{-}R)]^2$ must be replaced by $Q(0) \times Q(1)$

$[R/(1\text{-}R)]^3$ must be replaced by $Q(0) \times Q(1) \times Q(2)$

The details of the derivation are entirely straightforward but somewhat tedious. They will not be presented here. The correctness of equations (3-30) through (3-33) can be verified by substituting the algebraic expressions that these equations provide for P(0), P(1), P(2) and P(3) into balance equations (3-18) through (3-21).

As mentioned in Section 1.7, equations (3-30) through (3-33) can be used to assess the impact that deviations from the assumption of empirical independence have on the values of P(j). For example, suppose the values of R(j) are not all exactly equal to one another, but instead fall within a reasonably tight range. In particular, assume that $R(0) = .64$, $R(1) = .60$, $R(2) = .63$ and $R(3) = .61$. In such cases practitioners are interested in understanding how much error is introduced by ignoring these relatively small deviations and simply substituting the overall value of R into equations (3-26) through (3-29) when predicting the values of P(j).

Begin by noting that equations (3-30) through (3-33) can be used to determine the exact values of P(j) in this case.   When $R(0) = .64$, $R(1) = .60$, $R(2) = .63$ and $R(3) = .61$, the exact values of P(j) are given by equations (3-34) through (3-37).

$$P(0) = .107 \qquad\qquad (3\text{-}34)$$

$$P(1) = .170 \qquad\qquad (3\text{-}35)$$

$$P(2) = .276 \qquad\qquad (3\text{-}36)$$

$$P(3) = .447 \qquad\qquad (3\text{-}37)$$

Once the values of P(j) are determined, the overall value of R, which is defined by equation (3-4), can be computed using equation (3-38).

$$R = P(0) \times R(0) + P(1) \times R(1) + P(2) \times R(2) + P(3) \times R(3) \qquad (3\text{-}38)$$

Note that this value of R corresponds to the overall value an observer would actually measure in practice. It is valid for all possible values of R(0), R(1), R(2) and R(3), and reduces immediately to the trivial equation R = R when all four values of R(j) are the same.

For the values of R(j) specified in equations (3-34) through (3-37), this equation yields:

$$R = .617 \qquad\qquad (3\text{-}39)$$

Now that the overall value of R has been determined, it is possible to substitute this value into equations (3-26) through (3-29) to compute the corresponding values of P(j). This computation is, of course, based on the incorrect assumption that all four values of R(j) are equal to R. Equations (3-40) through (3-43) present the values of P(j) computed in this case.

$$P(0) = .107 \tag{3-40}$$

$$P(1) = .172 \tag{3-41}$$

$$P(2) = .276 \tag{3-42}$$

$$P(3) = .445 \tag{3-43}$$

Practitioners will be interested to compare the exactly correct values of P(j) in equations (3-34) through (3-37) with the approximately correct values in equations (3-40) through (3-43). The error introduced by the assumption of empirical independence is quite minor in this particular case, which explains why the solution presented in equations (3-26) through (3-29) and (1-1) through (1-4) can work so well in practice. Other cases can be analyzed similarly.

Equations (3-30) through (3-33) are also of interest in cases where the value of each individual R(j) can be estimated with a high degree of confidence. As discussed in Section 7.10.1, these estimates can sometimes be obtained in queuing network models by making the assumption that online behavior is equal to off-line behavior (Denning and Buzen 1978).

3.6 Extensions and generalizations

The random walk discussed in Chapters 1, 2 and 3 provides a specific example that illustrates issues, principles and techniques that are broadly applicable. The remaining chapters of this book provide a more general discussion of this entire area. In particular, Chapters 4 and 5 extend the discussion to the class of problems that are traditionally solved by deriving steady state distributions of discrete time Markov chains. Chapters 6 and 7 then extend the discussion further to Markov processes that operate in continuous time.

The following points, which have already been considered in the context of random walks, remain fundamentally important in all these cases.

1. The observable properties of systems operating over intervals of time are characterized by trajectories.

2. The rules that govern the dynamic operation of systems are characterized by deterministic state transition diagrams. These diagrams impose structural constraints on the associated trajectories.

3. Formal variables represent observable quantities that are defined in terms of trajectories. These definitions are based on the measurement procedures that practitioners routinely employ when they determine the actual values of these quantities.

4. For each trajectory that lies within the scope of the analysis, the exact values of all observable quantities are assumed to be known. However, most of these exact values are immaterial to the analysis. Exact values will vary from one trajectory to the next, but the derived results will be valid in all cases where the assumptions of the analysis are satisfied. This particular aspect of immateriality enables the analysis to accommodate uncertainty without invoking the sampling premise and assuming that observable quantities are samples drawn from probability distributions.

5. Analyses deal exclusively with trajectories that correspond to single contiguous intervals of time. This assumption implies that every transition into a state is followed directly by a transition out. As already noted, there are two exceptions to this general principle: the last transition into the final state is not followed by a

transition out, and the first transition out of the initial state is not preceded by a transition in. If the initial and final states are identical (i.e., matched), the total number of transitions into and out of each state must be equal.

6. The assumption of matched endpoints leads ultimately to sets of balance equations (e.g., equations (3-22) through (3-25)) that are direct counterparts of the equations that characterize the stationary (steady state) distributions of ergodic stochastic processes. This makes it possible to derive results within the framework of observational stochastics that are direct counterparts of certain results derived using the traditional theory of stochastic processes. Exact solutions can also be derived for cases where endpoints are not matched. These solutions are slightly more complex and have no direct counterparts in the traditional theory. See Section 4.7.2.

3.7 Formal characterization of observational stochastics

As noted in Section 2.3, random variables and stochastic processes are abstract entities that exist within the realm of pure mathematics. To achieve a comparable level of rigor, the intuitive concepts employed in observational stochastics must also be defined in formal mathematical terms. Practitioners who are comfortable with the intuitive notions of system, workload, trajectory and observable quantity, and who have a limited interest in mathematical formalisms, may wish to skip the material in the next few sections and proceed directly to Chapter 4.

3.7.1 State transition diagrams and finite state automata

On an intuitive level, state transition diagrams use simple graphical conventions to represent the step-by-step behavior of real or hypothetical systems. It is assumed that these systems can be in a number of different observable states. Each state is represented by a non-negative integer and is displayed within a circle as shown in

Figure 2-1. The total number of states must be finite.

System behavior is driven by a sequence of external events. Some of these external events cause the system to enter a new state. Others leave the current state unchanged. Each such transition is represented by a transition arrow that connects the current state to the state the system enters immediately following the corresponding external event. If the current state and the next state are the same, the transition arrow loops back to the current state. Such loops are shown in Figure 3-1 at states 0 and 3.

Formally speaking, states are simply mathematical abstractions labeled with the integers 0 through N. Their interpretation is entirely arbitrary and consistent with Kolmogorov's (1933) characterization of the states in a state space: "What the elements of this set represent is of no importance in the purely mathematical development of the theory."

Transition arrows are ordered pairs of states (j, k) where state j is connected to the tail of the arrow and state k is connected to its head. Each transition arrow is associated with one or more input symbols. Input symbols are mathematical abstractions whose interpretations are once again entirely arbitrary. On an intuitive level, input symbols correspond to external events.

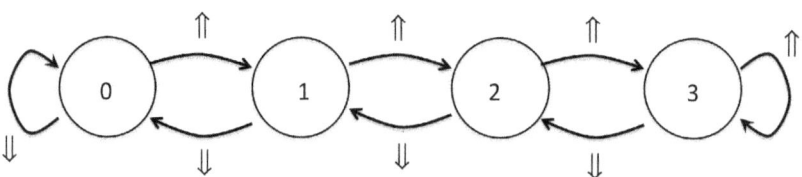

Figure 3-1. Formal state transition diagram for a random walk

The formal state transition diagram in Figure 3-1 is derived from the intuitive state transition diagram in Figure 2-1 by labeling the transition arrows with input symbols. In this particular example, there are two input symbols: ⇑ and ⇓. Informally, ⇑ corresponds to a right turn and ⇓ corresponds to a turn to the left.

Within the realm of pure mathematics, a state transition diagram is characterized formally by its transition function $F(j,\omega)$. The first argument j represents a state, and the second ω represents an input symbol. The value of $F(j,\omega)$ is also a state. Intuitively $F(j,\omega)$ represents the state that is entered after processing input symbol ω while in state j. The transition function $F(j,\omega)$ for the specific state transition diagram in Figure 3-1 is given by equations (3-44) through (3-51).

$$F(0, ⇓) = 0 \tag{3-44}$$

$$F(0, ⇑) = 1 \tag{3-45}$$

$$F(1, ⇓) = 0 \tag{3-46}$$

$$F(1, ⇑) = 2 \tag{3-47}$$

$$F(2, ⇓) = 1 \tag{3-48}$$

$$F(2, ⇑) = 3 \tag{3-49}$$

$$F(3, ⇓) = 2 \tag{3-50}$$

$$F(3, ⇑) = 3 \tag{3-51}$$

Readers with backgrounds in Computer Science will recognize immediately that the state transition diagram in Figure 3-1 is an example of a formal structure known as a Finite State Automaton or Finite State Machine (Mealy 1955). These abstract mechanisms are used in the specification and design of electronic circuits and in the analysis of algorithms. Their application to trajectories whose detailed properties are uncertain is a more recent development.

3.7.2 Workloads, trajectories and compound trajectories

A workload is an ordered string of input symbols. In observational stochastics, the number of possible input symbols in a workload is always finite. Figure 3-2 presents an example of a workload associated with the state transition diagram in Figure 3-1. The input symbols in this case are ⇑ and ⇓. As already noted, they correspond informally to right and left turns.

⇑⇑⇑⇑⇑⇑⇑⇑⇑⇑⇑⇑⇑⇑⇑⇑⇑⇑⇓⇑⇑⇓⇑⇓⇑⇓⇑⇓⇑⇓⇓⇓⇓⇑⇓⇑⇑⇓⇑⇓⇑⇓⇓⇓⇑⇓⇑⇑

Figure 3-2. Example of a workload

A trajectory is an ordered string of states: s(0), s(1), s(2) ... . Intuitively, a trajectory is generated when a workload is processed by a finite state automaton.

Note that trajectories can be represented in two alternative formats. The graphical format, which is illustrated in Figure 1-4, has already been discussed at length. The alternative format is simply a linear string of integers, where each integer represents a state. The linear string associated with the trajectory in Figure 1-4 is shown in Figure 3-3.

23333333333333333332323232323232321212121001012

Figure 3-3. Alternative representation of the trajectory in Figure 1-4

To specify a trajectory in a formal mathematical sense, it is necessary to define three separate entities: a finite state automaton, a workload $\omega(1)$, $\omega(2)$, $\omega(3)$ ... , and an initial state s(0). As already noted, a finite state automaton is specified by identifying its states, its input symbols and its transition function $F(j, \omega)$.

Once all these elements have been specified, the trajectory can be generated by successive iterations of equation (3-52). For example, the trajectory in Figure 3-3 is generated by the transition function specified by equations (3-44) through (3-51), the workload in Figure 3-2, and the initial state s(0) set equal to 2. This is somewhat similar to the way successive iterations of equations (2-9) through (2-12), in combination with an initial state, generate the sequence of random variables (i.e., the stochastic process) used in Chapter 2 as a model of a random walk (see Table 2-1).

$$s(j) = F(s(j-1), \omega(j)) \tag{3-52}$$

Observational stochastics is, in essence, the study of trajectories generated by finite state automata. In typical applications, the analyst knows the structure of the state transition diagram and its transition function $F(j,\omega)$. However, the step-by-step details of the workload being processed and the trajectory this workload generates remain uncertain and have no direct influence on the results of the analysis. Results derived through observational stochastics apply to all workloads and trajectories that satisfy the loose constraints (Section 3.7.5) used in the model being analyzed.

In some cases, it is necessary to introduce a closely related concept known as a compound trajectory. These structures are created by interleaving the states of a trajectory and the input symbols of a workload to form a single string. For example, the compound trajectory associated with the workload in Figure 3-2 and the trajectory in Figure 3-3 begins as follows:

2⇑3⇑3⇑3⇑3⇑3⇑3⇑3⇑3⇑ ...

Figure 3-4. Initial characters in a compound trajectory

3.7.3 Measurement of directly observable quantities

One of the underlying principles of observational stochastics is that all symbolic variables must represent quantities that are directly observable and measurable. The first step in formalizing this intuitive principle is to define the entities that are being observed.

In the case of observational stochastics, states and input symbols are regarded as basic observable quantities because they can be linked directly to physical entities that exist in the real world. Thus, trajectories, workloads and compound trajectories are also considered to be directly observable.

The next step is to describe the process by which observable quantities are measured. It is helpful to begin with a specific example. Recall that the observable quantity $R(2)^{\#}$ represents the number of times the walker exits from station 2 and turns to the right. As illustrated in Figure 3-1, this right turn will cause the walker to proceed from station 2 to station 3. Thus, $R(2)^{\#}$ can be measured by simply counting the number of times the pattern "23" appears within a trajectory. Measurement processes of this type play a central role in observational stochastics.

Formally, the process of measuring a directly observable quantity involves two steps:

Step A.   Define a *pattern* by specifying a string of characters. These characters correspond to states that appear in trajectories or input symbols that appear in workloads. In most cases, a pattern contains either 1 or 2 characters.

Step B. Count the number of times this pattern appears over the course of a trajectory, workload or compound trajectory. The counting process is defined formally by a simple pattern matching function MATCH(*pattern, string*) that returns the number of times *pattern* appears in *string*: e.g., MATCH(32, 2333210) is equal to 1. When patterns overlap, each instance is counted separately. Thus, MATCH(33, 2333210) is equal to 2.

In observational stochastics, the process of pattern matching and enumeration is the most fundamental of measurement activities. Symbolic variables that can be quantified through this basic process are identified with hash tags. Raw counts obtained in this manner can, of course, be combined arithmetically to quantify other directly observable variables. The wild card symbol * can be included in *pattern* to streamline this process.   For example, MATCH(*, *string*) is equal to the total length of s*tring*.

3.7.4 Extensions for continuous time processes

This formal characterization of measurement and observation must be extended when the flow of time is continuous rather than discrete. In such cases, all states that appear in a trajectory and all input symbols that appear in a workload must be assigned a second observable attribute: a timestamp that represents an associated time of occurrence. For the states in a trajectory, the time of occurrence is the instant each state is entered.   For the input symbols in a workload, it is the instant each input symbol arrives. In either case, the time values assigned to successive states in a trajectory or to successive input symbols in a workload must form a steadily increasing sequence: i.e., time must flow in a positive direction.

In cases of this type, it is often necessary to measure the amount of time a state is occupied over the course of a trajectory. This quantity can be measured by noting the time each state is entered (using the corresponding timestamp) and then subtracting that value from the time the next state in the trajectory is entered. The lengths of these individual intervals must then be summed over every instance in which the state appears. This entire process provides another example of direct measurement within observational stochastics. Although this straightforward approach to measuring intervals of time is sufficient for most cases of practical interest, it cannot be applied in certain limiting cases where the number of intervals that must be measured approaches infinity and the length of each interval approaches zero. These special cases, which are discussed further in Sections 8.6 and 9.2.6, can be safely ignored by most practitioners.

3.7.5 Loosely constrained deterministic (LCD) models

A model that combines a finite state automaton (i.e., a set of states, a set of input symbols, and a transition function) with a set of loose constraints on associated workloads and trajectories will be referred to as a loosely constrained deterministic (LCD) model (Buzen 2012).   From a formal mathematical perspective, observational stochastics can be regarded as the study of these LCD models. In contrast, the formal object of analysis in traditional stochastic modeling is a stochastic process: that is, a sequence of random variables having the same sample space but distinct probability distributions.

LCD models make it possible to carry out analyses where uncertainty exists, but where there is no compelling reason to invoke the sampling premise and assume that observable values can be regarded as samples that have been drawn from underlying probability distributions. The workload in Figure 3-2 illustrates just such a case. This workload, which begins with 17 right turns in a row, could have been generated by 45 independent tosses of a coin.   However, without external evidence, this assumption appears highly unlikely.

On the other hand, a practitioner who observes the trajectory in Figure 1-4 (or, equivalently, Figure 3-3) can easily measure the values of R(j). In this case, all four values of R(j) are equal to $^2/_3$. Thus, the assumption of empirical independence (a loose constraint) is satisfied. Since it is also true that the endpoints of the trajectory are matched, it is possible to conclude with certainty that equations (3-26) through (3-29) provide the correct values of P(j) for this trajectory. All other properties of the trajectory are immaterial and have no direct impact on the analysis. Definitive conclusions of this type are the hallmark of observational stochastics.

CHAPTER 4

Generalized Random Walks

4.1 Overview

The principles of observational stochastics extend well beyond the specific example presented in Chapter 3. In fact, observational stochastics is applicable in many situations where Markov models would otherwise be employed. This chapter builds upon the concepts and techniques presented earlier in order to develop these extensions.

Even though the overall flow of the analysis is quite similar to Chapter 3, the mathematical notation required to deal with the general case adds complexity to the discussion. Readers interested in the applications of observational stochastics rather than the mathematical equations that must be solved when dealing with the general case may wish to review the following seven points and then proceed directly to Chapter 5.

1. The techniques used in Chapter 3 to analyze the simple random walk can be extended to deal with random walks where the number of stations is equal to any value N and where each station can be reached from every other station in a single step.

2. In this very general setting, the objective once again is to determine the proportion of visits the walker makes to each station over the course of the trajectory. These observable proportions are represented symbolically as P(0), P(1), P(2) … P(N).

3. If the endpoints of a trajectory are matched, the desired values of P(j) are given by the solution to a set of balance equations obtained by setting the number of entrances into each station equal to the number of exits. These balance equations are direct generalizations of equations (3-18) through (3-21). The entire set

of balance equations are specified   using standard mathematical notation in equation (4-22).

4. If the endpoints of a trajectory are not matched, the solution to the balance equations still provides a reasonable approximation to the actual values of P(j). This is true in most, but not all, cases. In the limit as trajectory length increases, the error associated with this approximation usually approaches zero. When required, exact solutions for trajectories with unmatched endpoints can be obtained through the analysis presented in Section 4.7.2.

5. The form of the general algebraic solution - with or without matched endpoints - is algebraically complex and of limited interest. Practitioners are typically concerned with solutions to specific sets of balance equations that are associated with individual models such as those presented in Chapters 3 and 5.

6. Trajectories with matched endpoints are closely related to a new class of formal objects known as t-loops (trajectory loops). To construct a t-loop, simply merge the initial and final stations of a trajectory with matched endpoints. This transforms the original linear trajectory into a closed loop with no beginning and no end. Some interesting properties of t-loops are noted in section 4.7.4.

7. The general solution for the values of P(0), P(1), P(2) ... P(N) obtained using observational stochastics has exactly the same algebraic form as the steady state distribution of the corresponding Markov chain.   However, in a traditional time-homogeneous Markov chain, the transition probabilities that control step-by-step behavior are the same for every step of an associated trajectory. In contrast, the global transition matrices employed in observational stochastics are comprised of overall values derived from complete trajectories.   The detailed step-by-step mechanisms that regulate these trajectories have no impact on the analysis and remain immaterial.   This enables derived results to apply more broadly.

This issue is discussed in general terms in Section 4.8 and also revisited in Section 5.6 of Chapter 5.

4.2 Generalized walks: 4 stations

Begin with the simple random walk whose state transition diagram is illustrated in Figure 2-1. A modified version of this diagram appears below as Figure 4-1. The circles representing the four stations are now arranged to form a square rather than a straight line. Apart from this minor change, all other aspects of the two diagrams are identical.

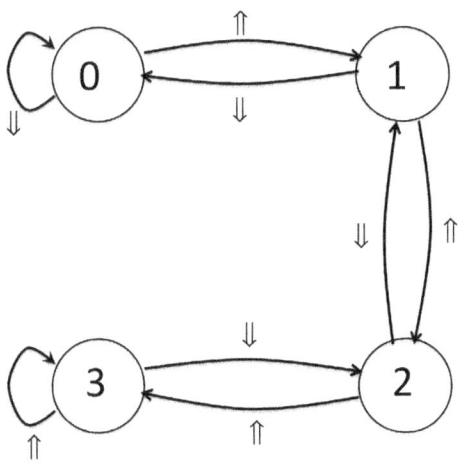

Figure 4-1. State transition diagram for random walk

Now consider the more general diagram illustrated in Figure 4-2. The four stations in this diagram are once again arranged to form a square. However, every station is now connected to every other station by a transition arrow. Intuitively, this means it is possible for the walker to reach every station from every other station in a single transition.

To simplify Figure 4-2, the input symbols associated with each transition have been omitted. These input symbols play no direct role in this particular analysis and need not be specified.

As in the case of the simple random walk, a trajectory is again defined as a sequence of integers belonging to the set {0,1,2,3}. The integers correspond to the stations a walker visits during a particular walk. The goal of the analysis is to express the proportion of visits the walker makes to each station (i.e., P(0), P(1), P(2) and P(3)) as a function of the selections the walker makes when exiting from each station.

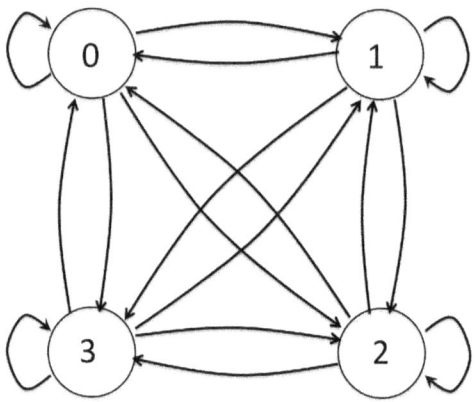

Figure 4-2. Generalized state transition diagram

In Chapter 3, the proportions P(0), P(1), P(2) and P(3) were shown to satisfy a set of balance equations: specifically, equations (3-18) through (3-21). The goal of the next few sections is to develop a comparable set of balance equations that can be solved to obtain the desired values of P(0), P(1), P(2) and P(3).

The values R(0), R(1), R(2) and R(3) that appear on the right hand sides of balance equations (3-18) through (3-21) represent the proportion of turns the walker makes to the right after leaving stations 0, 1, 2 and 3 respectively. To extend the analysis in Chapter 3 to the general class of trajectories associated with Figure 4-2, begin by noting that the walker now has four rather than two options when exiting from station 0, 1, 2 and 3. The proportion of time the walker selects each option can be associated with a separate observable quantity. For example, consider a walker currently located at station 0. The next transition can take the walker to stations 0, 1, 2 or 3, and the proportion of time each of these possible transitions occur over the course of an entire trajectory can be represented as follows:

$X(0,k)$ = proportion of exits from station 0 that are followed immediately by a visit to station k (for k = 0, 1, 2, 3)

= proportion of appearances of the integer 0 that are followed immediately by the integer k in the trajectory being analyzed

The same definitions can be extended to stations 1, 2 and 3. In all four cases, the observable variables $X(j,k)$ will be defined as follows:

$X(j,k)$ = proportion of exits from station j that are followed immediately by a visit to station k (defined for j = 0, 1, 2, 3 and for k = 0, 1, 2, 3).

= proportion of appearances of the integer j that are followed immediately by the integer k in the trajectory being analyzed

The sixteen values of $X(j,k)$ can be arranged to form a square array as shown in Figure 4-3. Arrays of this type will be referred to as global transition matrices because the values that appears in these arrays summarize overall properties of the trajectory as a whole. As discussed in Section 4.8, these matrices differ from the stepwise transition matrices that characterize the step-by-step behavior of traditional Markov chains.

$$\begin{pmatrix} X(0,0) & X(0,1) & X(0,2) & X(0,3) \\ X(1,0) & X(1,1) & X(1,2) & X(1,3) \\ X(2,0) & X(2,1) & X(2,2) & X(2,3) \\ X(3,0) & X(3,1) & X(3,2) & X(3,3) \end{pmatrix}$$

Figure 4-3. Global transition matrix – general case

The corresponding global transition matrix for the simple random walk in Section 3.4 is shown in Figure 4-4. Note that many of the entries in this matrix are zero. These zeros correspond to transitions that cannot possibly occur because they are excluded from the state transition diagram in Figure 2-1 and Figure 4-1.

$$\begin{pmatrix} R & 1-R & 0 & 0 \\ R & 0 & 1-R & 0 \\ 0 & R & 0 & 1-R \\ 0 & 0 & R & 1-R \end{pmatrix}$$

Figure 4-4. Global transition matrix – random walk

4.3 Balance equations

Despite the totally flexible structure of the generalized random walk in Figure 4-2, one critically important point remains unchanged: it is still true that each entry the walker makes into a station is followed by an exit out of that station. Thus, for each station, the total number of entries and exits must be the same. There are, as before, two minor exceptions to this general rule. The final station has one extra entrance because the last entrance into the final station is not followed by an exit. Similarly, the initial station has one extra exit because the first exit from the initial station is not preceded by an entry. If the initial and final stations are the same (i.e., if the trajectory has matched endpoints), these two exceptions will cancel each other. In such cases the total number of entries and exits will be the same for each of the four stations.

The analysis presented in Chapter 3 can now be extended to this more general setting. Recall that the values $R(j)^{\#}$ and $L(j)^{\#}$ were defined in Chapter 3 as follows:

$R(j)^{\#}$ = the number of times the walker leaves station j and turns right (for j = 0, 1, 2, 3)

$L(j)^{\#}$ = the number of times the walker leaves station j and turns left (for j = 0, 1, 2, 3)

These definitions can be generalized as follows:

$X(j,k)^{\#}$ = the number of times the walker leaves station j and proceeds next to station k

(for j = 0, 1, 2, 3 and k = 0, 1, 2, 3)

As in Chapter 3, the hash tag incorporated into the symbol $X(j,k)^{\#}$ highlights the fact that these variables represent basic measurements (i.e., raw counts). As specified in Section 3.7.3, the measurement process used to evaluate these observable quantities employs the function MATCH(*jk*, *string*) where *string* is the entire trajectory.

Since the walker can proceed directly to any of the four stations after leaving station j, the total number of times the walker exits from station j is:

$$X(j,0)^{\#} + X(j,1)^{\#} + X(j,2)^{\#} + X(j,3)^{\#} = \sum_{k=0}^{3} X(j,k)^{\#}$$

Similarly, since the walker can arrive at station j after exiting from any of the four stations, the total number of times the walker enters station j is:

$$X(0,j)^{\#} + X(1,j)^{\#} + X(2,j)^{\#} + X(3,j)^{\#} = \sum_{k=0}^{3} X(k,j)^{\#}$$

Since the total number of exits from station j is always equal to the total number of entrances into station j when a trajectory's endpoints are matched, the values of $X(j,k)^{\#}$ must satisfy equations (4-1) through (4-4).

$$X(0,0)^{\#} + X(0,1)^{\#} + X(0,2)^{\#} + X(0,3)^{\#} =$$

$$X(0,0)^{\#} + X(1,0)^{\#} + X(2,0)^{\#} + X(3,0)^{\#} \qquad (4\text{-}1)$$

$$X(1,0)^{\#} + X(1,1)^{\#} + X(1,2)^{\#} + X(1,3)^{\#} =$$

$$X(0,1)^{\#} + X(1,1)^{\#} + X(2,1)^{\#} + X(3,1)^{\#} \qquad (4\text{-}2)$$

$$X(2,0)^{\#} + X(2,1)^{\#} + X(2,2)^{\#} + X(2,3)^{\#} =$$

$$X(0,2)^{\#} + X(1,2)^{\#} + X(2,2)^{\#} + X(3,2)^{\#} \qquad (4\text{-}3)$$

$$X(3,0)^{\#} + X(3,1)^{\#} + X(3,2)^{\#} + X(3,3)^{\#} =$$

$$X(0,3)^{\#} + X(1,3)^{\#} + X(2,3)^{\#} + X(3,3)^{\#} \qquad (4\text{-}4)$$

Using standard mathematical notation, equations (4-1) - (4-4) can be written more compactly as equation (4-5):

$$\sum_{k=0}^{3} X(j,k)^{\#} = \sum_{k=0}^{3} X(k,j)^{\#} \qquad \text{for } j = 0, 1, 2, 3 \qquad (4\text{-}5)$$

Equations (4-1) through (4-4) are direct generalizations of equations (3-6) through (3-9) in Chapter 3. The next step is to transform these equations into a set of balance equations that can be solved to obtain the desired values of P(0), P(1), P(2) and P(3). This process begins with a formal definition of the symbolic variable X(j,k) in the global transition matrix shown in Figure 4-3.

X(j,k) = proportion of time the walker proceeds to station k after leaving station j (as shown in Figure 4-3).

$$X(j,k) = \frac{X(j,k)^{\#}}{X(j,0)^{\#} + X(j,1)^{\#} + X(j,2)^{\#} + X(j,3)^{\#}} \qquad (4\text{-}6)$$

$$\text{for } j, k = 0, 1, 2, 3$$

It follows immediately from equation (4-6) that:

$$X(j,0) + X(j,1) + X(j,2) + X(j,3) = 1 \qquad (4\text{-}7)$$

$$\text{for } j = 0, 1, 2, 3$$

Note that equation (4-6) generalizes the definition of R(j) presented in equation (3-1). The definitions of V(j), V and P(j), which appear in equations (3-2), (3-3) and (3-5), can also be generalized as follows:

V(j) = total number of visits the walker makes to station j (if station j is the final station in the trajectory, the last visit is not counted)

= total number of exits the walker makes from station j

$$= X(j,0)^{\#} + X(j,1)^{\#} + X(j,2)^{\#} + X(j,3)^{\#} \qquad (4\text{-}8)$$
$$\text{for } j = 0, 1, 2, 3$$

V   = total number of visits the walker makes to all stations

   (excluding the last visit to the final station in the trajectory)

$$= V(0) + V(1) + V(2) + V(3) \qquad (4\text{-}9)$$

$P(j)$ = proportion of visits to station j

$$= V(j) / V \qquad\qquad \text{for } j = 0, 1, 2, 3 \qquad (4\text{-}10)$$

It follows immediately from equation (4-10) that:

$$P(0) + P(1) + P(2) + P(3) = 1 \qquad (4\text{-}11)$$

Equation (4-11) expresses the same normalization constraint mentioned in the discussion following equation (2-23).   This constraint is used throughout Chapters 2 and 3, and will be used in this chapter as well.

Equations (4-6) and (4-8) imply

$$X(j,k)^{\#} = V(j) \times X(j,k)$$

$$= V \times P(j) \times X(j,k) \qquad (4\text{-}12)$$

Equations (4-8) and (4-10) imply

$$X(j,0)^{\#} + X(j,1)^{\#} + X(j,2)^{\#} + X(j,3)^{\#} = V(j)$$

$$= V \times P(j) \qquad (4\text{-}13)$$

Equations (4-1) through (4-4) can now be simplified as follows. Equation (4-13) implies $X(j,0)^{\#} + X(j,1)^{\#} + X(j,2)^{\#} + X(j,3)^{\#}$ can be replaced by $V \times P(j)$ on the left side of each equation. In addition, equation (4-12) implies each occurrence of $X(j,k)^{\#}$ on the right side of these four equations can be replaced by $V \times P(j) \times X(j,k)$.

After making these substitutions, both sides of each simplified equation can be divided by V to yield the following set of balance equations:

$$P(0) = P(0) \times X(0,0) + P(1) \times X(1,0) + P(2) \times X(2,0) + P(3) \times X(3,0)$$
(4-14)

$$P(1) = P(0) \times X(0,1) + P(1) \times X(1,1) + P(2) \times X(2,1) + P(3) \times X(3,1)$$
(4-15)

$$P(2) = P(0) \times X(0,2) + P(1) \times X(1,2) + P(2) \times X(2,2) + P(3) \times X(3,2)$$
(4-16)

$$P(3) = P(0) \times X(0,3) + P(1) \times X(1,3) + P(2) \times X(2,3) + P(3) \times X(3,3)$$
(4-17)

Note that these balance equations can be written more compactly as follows:

$$P(j) = \sum_{k=0}^{3} P(k) \times X(k,j) \quad \text{for } j=0,1,2,3$$
(4-18)

[Minor comment: as one would expect, equations (4-14) through (4-17) reduce immediately to the balance equations for a random walk (i.e., equations (3-18) through (3-21)) when each occurrence of $X(j,k)$ is replaced by the corresponding entry in the global transition matrix that is shown in Figure 4-4.]

4.4 Solving the balance equations

The solution to equations (4-14) through (4-17) is a routine but tedious exercise in linear algebra. The key steps are presented here. These details are not required to follow the main thread of the discussion, which resumes in Section 4.5.

1. Use equation (4-17) to express P(3) as a function of P(0), P(1) and P(2). This yields:

$$P(3) = \frac{P(0) \times X(0,3) + P(1) \times X(1,3) + P(2) \times X(2,3)}{1 - X(3,3)}$$

$$= P(0) \times Z(0,3) + P(1) \times Z(1,3) + P(2) \times Z(2,3) \qquad (4\text{-}19)$$

where

$$Z(j,3) = \frac{X(j,3)}{1 - X(3,3)} \qquad \text{for } j = 0, 1 \text{ and } 2$$

2. Replace each occurrence of P(3) in equations (4-14) through (4-16) by the expression derived in Step 1. This eliminates P(3) from these three equations.

3. Use the modified version of equation (4-16) to express P(2) as a function of P(0) and P(1). This yields:

$$P(2) = \frac{P(0) \times [X(0,2) + Z(0,3)] + P(1) \times [X(1,2) + Z(1,3)]}{1 - X(2,2) - Z(2,3)}$$

$$(4\text{-}20)$$

4. Replace each occurrence of P(2) in equations (4-14), (4-15) and (4-19) by the expression derived in Step 3. This eliminates P(2) from these three equations.

5. Use the twice modified version of equation (4-15) to express P(1) as a function of P(0).   This function is significantly more complex than equation (4-18) and need not be shown here.

6. Replace each occurrence of P(1) in equations (4-14), (4-19) and (4-20) by the expression derived in Step 5. This eliminates P(1) from these three equations. Thus, P(1), P(2) and P(3) are now expressed as functions of P(0). In addition, equation (4-14) is left with only one remaining unknown: P(0).

Under normal circumstances, it would be possible to use equation (4-14) to derive an expression for P(0) in terms of the X(j,k). However, because $X(j,0)+X(j,1)+X(j,2)+X(j,3)=1$, equations (4-14) through (4-17) are not all independent. As a result, the version of equation (4-14) generated by Step 6 ultimately reduces to:

$$P(0) = P(0)$$

7. Since equation (4-14) has become a trivial identity, it cannot be used to obtain an expression for P(0). An additional independent equation is thus required to complete the analysis. As usual, this additional equation is provided by a normalization constraint: in this case, equation (4-11).

Simply replace the variables P(1), P(2) and P(3) that appear in equation (4-11) with functions of P(0) obtained in Step 6. Once these substitutions have been made, an explicit algebraic expression for P(0) can be derived. Expressions for the other values of P(j) can then be obtained using the results of Step 6.

This process leads to closed form algebraic expressions for P(0), P(1), P(2) and P(3) that are highly complex and enormously cumbersome to employ. Because of their unwieldy nature, practitioners are seldom interested in highly generalized solutions of this type. Instead, practitioners are more interested in compact algebraic expressions for P(j) that are applicable to specific models such as the random walk discussed in Chapter 3.

For each specific model, the detailed structure of the state transition diagram will imply that certain transitions are impossible. The values of X(j,k) associated with these transitions will, of course, be set to zero in the model's global transition matrix.   If the remaining non-zero values of X(j,k) are then expressed as simple functions of the model's observable parameters (e.g., R), it is sometimes possible to obtain compact closed form solutions such as those presented in equations (3-22) through (3-25) and in equations (3-30) through (3-33). Obtaining such solutions is a major goal of both observational stochastics and traditional stochastic modeling.

4.5 Balance equations and eigenvectors

Readers familiar with matrix algebra will recognize immediately that equation (4-18) implies that [P(0), P(1), P(2), P(3)] is an eigenvector of the global transition matrix [X(j,k)], and that the associated eigenvalue is equal to 1.   In ordinary language, this simply means that multiplying the vector [P(0), P(1), P(2), P(3)] by the matrix [X(j,k)] generates a new vector that is exactly equal to the original vector [P(0), P(1), P(2), P(3)].

Eigenvectors and eigenvalues are objects of intense mathematical interest and genuine importance in the real world. One well known result in this area has a direct bearing on this discussion: if the global transition matrix in Figure 4-3 is associated with a trajectory whose endpoints are matched, that matrix is guaranteed to have one and only one eigenvector that satisfies the normalization constraint $P(0) + P(1) + P(2) + P(3) = 1$. In other words, the general solution procedure described in Section 4.4 always yields a unique solution for the values of P(j). The formal proof of this result is beyond the scope of this discussion; however, readers with advanced backgrounds should note that a trajectory with matched endpoints always generates a global transition matrix that has a single irreducible subchain. Such matrices are sometimes referred to as regular transition matrices.

4.6 Generalized walks: N+1 stations

It is a routine matter to generalize these findings to trajectories where the number of states is arbitrary and where each state can be reached from every other state in a single transition. This section presents a formal analysis of this completely general case.

Consider any trajectory of length L+1 with N+1 possible states (i.e., consider any string of integers s(0), s(1), s(2) ... s(L) where $0 \leq s(j) \leq N$ for each value of j).

As in Section 4.3, define $X(j,k)^{\#}$ as the total number of transitions from state j to state k for this trajectory (i.e., $X(j,k)^{\#}$ is equal to MATCH(jk, *trajectory*), the number of times that j is followed immediately by k in the string of integers that forms the trajectory).

Lemma 4.1

If $s(0) = s(L)$, the values of $X(j,k)^{\#}$ satisfy equation (4-21).

$$\sum_{k=0}^{N} X(j,k)^{\#} = \sum_{k=0}^{N} X(k,j)^{\#} \qquad \text{for } j = 0, 1, 2 ...N \qquad (4\text{-}21)$$

Note that the left hand side of equation (4-21) represents the total number of transitions out of state j. Similarly, the right hand side of equation (4-21) represents the total number of transitions into state j. Lemma 4.1 asserts that these two totals must always be equal to each other for all values of j, provided the endpoints of the trajectory are matched.

Proof:

Consider the cumulative number of transitions into and out of each state at any point along the trajectory. If these two cumulative totals are equal for a particular state at a given point, that state will be regarded as being *balanced* at that point. If not, the state will be regarded as being *imbalanced.* The proof of Lemma 4.1 depends on the shifting pattern of balanced and imbalanced states at each point along the trajectory.

Begin by defining a *transition* as an ordered pair of states within a trajectory $s(0)$, $s(1)$, $s(2)$ ... $s(L)$. The first transition in this trajectory is $[s(0), s(1)]$, the second is $[s(1), s(2)]$, and so on. The final transition is clearly $[s(L-1), s(L)]$,

Before the first transition, all states are balanced because the number of transitions into and out of every state is equal to zero. Assuming $s(1)$ is not equal to $s(0)$, the first transition $[s(0), s(1)]$ generates imbalances at two states: $s(0)$ and $s(1)$. Temporarily put aside the imbalance at state $s(0)$ and consider only the imbalance created by the transition into state $s(1)$.

Now consider the effect of the next transition $[s(1), s(2)]$. This transition corrects the imbalance at state $s(1)$ but also creates an imbalance at state $s(2)$. The process of correcting an existing imbalance and creating a new one continues with each subsequent transition. In general, the transition $[s(i-1),s(i)]$ corrects the imbalance at state $s(i-1)$ and creates an imbalance at state $s(i)$.

After completing all L transitions, only two imbalances remain: the imbalance at state $s(L)$ created by the last transition into that state, and also the original imbalance at state $s(0)$ created by the initial transition out of state $s(0)$ – and temporarily put aside. If $s(0) = s(L)$, these two imbalances will cancel each other and leave all N+1 states balanced. Equation (4-21) simply states that this balance condition exists for all states. This completes the proof for cases where $s(1)$ differs from $s(0)$.

Note that multiple replications of $s(0)$ at the start of the trajectory represent self loops that leave state $s(0)$ balanced. The presence of such self loops simply shifts the start of this proof to the first transition [s(i-1), s(i)] for which s(i) is different from s(0).

Lemma 4.1 can be used to extend equation (4-18) to more general trajectories having length L+1 and N+1 states. Before this can be done, it is necessary to extend the scope of equations (4-6), (4-8), (4-9) and (4-10) to accommodate trajectories comprised of N+1 states. These straightforward extensions are:

$$X(j,k) = X(j,k)^{\#} / \sum_{n=0}^{N} X(j,n)^{\#} \qquad \text{for } j, k = 0, 1, 2 \ldots N$$

X(j,k) is the proportion of transitions out of state j that lead directly to state k.

$$V(j) = \sum_{n=0}^{N} X(j,n)^{\#} \qquad \text{for } j = 0, 1, 2 \ldots N$$

V(j) is the total number of visits to state j over the course of the trajectory, excluding the final visit to state s(L) at the end of the trajectory.

$$V = \sum_{j=0}^{N} V(j)$$

V is the total number of visits to all states, excluding the final visit to state s(L) at the end of the trajectory. Note that V is equal to L, the number of transitions in the trajectory.

$$P(j) = V(j) / V \qquad \text{for } j = 0, 1, 2 \ldots N$$

P(j) is the proportion of visits that are made to state j. Note that the final visit to state s(L) is excluded from both the numerator and the denominator in the expression for P(j).

Theorem 4.1

Consider any trajectory of length L+1 with N+1 possible states (i.e., consider any string of integers s(0), s(1), s(2) ... s(L) where 0≤s(j)≤N for each value of j).   Assume that each state appears at least once in the trajectory so that the values of X(j,k) are all well defined.

If $s(0) = s(L)$, the values of P(j) are given by the normalized solution to the set of N+1 linear equations specified by equation (4-22).   In other words, the values of P(j) form the normalized eigenvector of the matrix [X(j,k)] shown in Figure 4-5.

$$P(j) = \sum_{k=0}^{N} P(k) \times X(k,j) \qquad \text{for } j = 0, 1, 2 \ldots N \qquad (4\text{-}22)$$

Proof:

The algebraic substitutions that lead from equation (4-5) to equation (4-18) can be combined with Lemma 4.1 to transform equation (4-21) into equation (4-22).

Begin with equation (4-21). Note that the left hand side is equal to V(j), which is equal to $V \times P(j)$. Also each occurrence of $X(k,j)^{\#}$ on the right side of equation (4-21) is equal to $V \times P(k) \times X(k,j)$.

After making these substitutions on both the left and right sides of equation (4-21), divide both sides by V. These straightforward steps transform equation (4-21) into equation (4-22), thus completing the proof.

Note that a solution to the N+1 linear equations specified by equation (4-22) is guaranteed to exist because a set of values of P(j) that satisfy these equations can be extracted directly from the trajectory upon which the values of X(j,k) are based.   As noted in

Section 4.5, the uniqueness of the normalized solution is assured because the matrix in Figure 4-5 has a single irreducible subchain.

$$\begin{pmatrix} X(0,0) & X(0,1) & X(0,2) & ... & X(0,N) \\ X(1,0) & X(1,1) & X(1,2) & ... & X(1,N) \\ \vdots & \vdots & \vdots & & \vdots \\ \vdots & \vdots & \vdots & & \vdots \\ X(N,0) & X(N,1) & X(N,2) & ... & X(N,N) \end{pmatrix}$$

Figure 4-5. Global transition matrix – N+1 states

4.7 Unmatched endpoints

The assumption of matched endpoints would seem to impose an overly restrictive constraint on the trajectories to which it applies. In particular, when the number of states in the state transition diagram is large, it would seem highly unlikely for the final state of a trajectory to be identical to its initial state.

From a mathematical perspective, the assumption of matched endpoints is merely a technical convenience. Its main purpose is to simplify the algebraic solution that is ultimately derived. As shown in Section 4.7.2, it is entirely feasible to carry out an exact analysis in cases where endpoints are not matched.

In practice, the error introduced by the assumption of matched endpoints is often very small, especially when the trajectory is long. In other words, the approximate solution obtained by assuming matched endpoints is typically very close to the exact solution obtained for cases where endpoints are not matched. The next two sections expand upon these points.

4.7.1 Intuitive expectations for long trajectories

Consider any trajectory with unmatched endpoints. Suppose this trajectory is partitioned into two segments: the first begins at the start of the trajectory and continues through the final appearance of the initial state; the second begins immediately after the final appearance of the initial state and continues to the end of the trajectory.

The first segment clearly corresponds to a trajectory with matched endpoints. The values of P(j) for this segment are, as usual, given by the normalized eigenvector of the corresponding global transition matrix. This solution is exactly correct.

The overall values of P(j) for the entire trajectory are the weighted average of the values of the P(j) for each of the two segments. If the second segment is substantially shorter than the first, it will have much less impact on the overall values of P(j).

In most cases, the second segment is likely to be substantially shorter than the first, provided the trajectory is long. To see why, note that the initial station is likely to reappear from time to time as the random walk proceeds. This implies that the length of the first segment will grow steadily as the overall length of the walk increases. However, the length of the second segment is not likely to increase because each successive reappearance of the initial state resets the length of the second segment to zero. This implies the values of P(j) and X(j,k) for the entire trajectory will be dominated by the values from the first segment, provided the trajectory is long. Thus the error introduced by the assumption of matched endpoints will become vanishingly small.

It should be noted, however, that there are certain cases where these informal expectations are not valid. As an example, consider the modified random walk whose state transition diagram is illustrated in Figure 4-6.   As this diagram indicates, a walker starting initially at station 2 can make an arbitrary number of visits to stations 2 and 3 before reaching station 1.   However, once the walker arrives at station 1, it is impossible to return to the initial station (i.e., station 2). In effect, a walker turning right after leaving station 1 encounters a new reflecting barrier and returns immediately to station 1. Using the traditional terminology of Markov chains, states 2 and 3 are transient. In contrast, states 0 and 1, which can continue to reappear from time to time as the trajectory grows, are said to be recurrent.

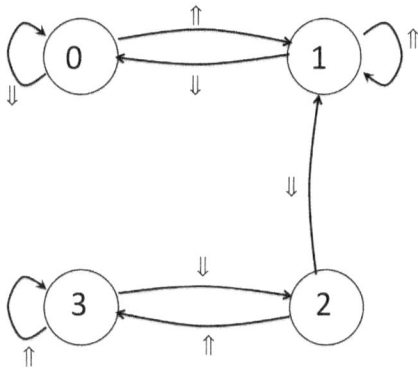

Figure 4-6. Transient states

If the initial state is transient, it is no longer reasonable to assume that this state will continue to reappear from time to time as the random walk proceeds. Instead, the length of the first segment will stabilize while the length of the second segment continues to increase. The arguments that have just presented in this section are not applicable to these very special cases.

4.7.2 Formal analysis

The exact analysis of trajectories with unmatched endpoints is based on an important relationship between two closely related trajectories. Begin by considering any initial trajectory whose endpoints are not matched. As usual, let the matrix shown in Figure 4-5 be the global transition matrix for this trajectory, and let P(j) represent the proportion of times state j appears over the entire trajectory. Following the conventions used throughout this book, the last state s(L) is excluded from the computation of each P(j).

The objective of this analysis is to derive algebraic expressions for the values of P(j). Since the endpoints are not matched, the desired values of P(j) do not correspond exactly to the normalized eigenvector of the global transition matrix in Figure 4-5.

Corollary 4.1

If the endpoints of a trajectory are not matched, the values of P(j) are given by the normalized solution to the set of N+1 linear equations specified by equation (4-23). In these equations, Y(j,k) is the proportion of transitions out of state j that lead directly into state k in an adjusted trajectory that is formed by replacing the final state s(L) in the original trajectory by the initial state s(0).

$$P(j) = \sum_{k=0}^{N} P(k) \times Y(k,j)$$

$$(4\text{-}23)$$

Proof:

By Theorem 4.1, the normalized solution to the set of N+1 linear equations specified by equation (4-23) yields the correct values of P(j) for the adjusted trajectory.

Recall that the final state s(L) is excluded from the computation of the values of P(j) for both the original trajectory and the adjusted trajectory. Since the two trajectories are identical except for their final states, the values of P(j) obtained by solving equation (4-23) must also represent the correct values of P(j) for the original trajectory. This completes the proof.

To apply Corollary 4.1 in practice, it is necessary to express the values of Y(j,k) for the adjusted trajectory in terms of the observable values of X(j,k) in the original trajectory.   With only two exceptions, each value of Y(j,k) is equal to the corresponding observable value of X(j,k). The two exceptions depend on the initial state in the trajectory s(0), the final state in the trajectory s(L), and also the next to the last state in the trajectory s(L-1).

Suppose $s(0) = a$, $s(L-1) = b$ and $s(L) = c$. Changing the final state in the original trajectory from state c to state a only affects two values in the matrix [X(j,k)] shown in Figure 4-5: the value of $X(b,c)$ must be reduced slightly to reflect the fact that $X(b,c)^{\#}$ has been reduced by one, and the value of X(b,a) must be increased slightly to reflect the fact that $X(b,a)^{\#}$ has been increased by one.

Note that the values of V(j) are the same in both the original and the adjusted trajectories. In particular, both trajectories have the same value of V(b). Thus, reducing the number of transitions from state b to state c by one implies that $Y(b,c)^{\#}$ is equal to $X(b,c)^{\#}$ minus 1. Dividing both sides by V(b) yields:

$$Y(b,c) = X(b,c) - \frac{1}{V(b)} \qquad (4\text{-}24)$$

Similarly, increasing the number of transitions from state b to state a by one implies that $Y(b,a)^{\#}$ is equal to $X(b,a)^{\#}$ plus 1. In this case, dividing both sides by V(b) yields:

$$Y(b,a) = X(b,a) + \frac{1}{V(b)} \qquad (4\text{-}25)$$

With the exception of Y(b,c) and Y(b,a), all other values of Y(j,k) that appear in equation (4-23) are equal to the corresponding values of X(j,k). Thus, the normalized eigenvector of the matrix shown in Figure 4-7, with Y(b,c) and Y(b,a) defined by equations (4-24) and (4-25), yields the correct values of P(j) for both the adjusted trajectory and the original trajectory.

$$\begin{pmatrix} X(0,0) & X(0,1) & & \cdots & & X(0,N) \\ X(1,0) & & & \cdots & & X(1,N) \\ \vdots & & & & & \vdots \\ \vdots & & & & & \vdots \\ X(b,0) & \cdots & Y(b,a) & \cdots & Y(b,c) & X(b,N) \\ \vdots & & & & & \vdots \\ \vdots & & & & & \vdots \\ X(N,0) & X(N,1) & & \cdots & & X(N,N) \end{pmatrix}$$

Figure 4-7. Global transition matrix for the adjusted trajectory

4.7.3 Example of limiting behavior

It is clear from equations (4-24) and (4-25) that $Y(b,c)$ will approach $X(b,c)$ and $Y(b,a)$ will approach $X(b,a)$ as $V(b)$ increases. In most cases of practical interest, this implies that the approximate values of $P(j)$ that are obtained from the matrix in Figure 4-5 will approach the exactly correct values of $P(j)$ that are obtained from the matrix in Figure 4-7 as the length of the trajectory increases. This section presents a simple deterministic example that illustrates the rate at which convergence occurs in a specific case. Readers with limited interest in these details may wish to proceed directly to Section 4.7.4.

The state transition diagram illustrated in Figure 4-8 represents a special type of random walk. There are three stations: 0, 1 and 2. The walker flips a coin to determine each successive move. If the coin comes up heads, the walker exits from the current station, encounters a reflecting barrier, and returns to that same station. If the coin comes up tails, the walker moves to the next station according to the following rule: if the walker is at station 0, the next station is station 1; if the walker is at station 1, the next station is station 2; if the walker is at station 2 , the next station is station 0. Assume also that this particular coin is imbalanced: heads are twice as likely to occur as tails.

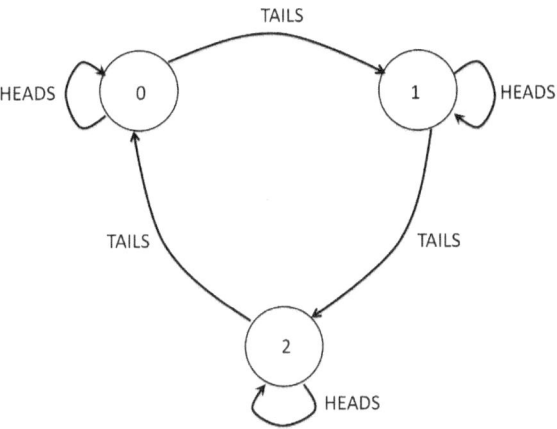

Figure 4-8. State transition diagram

Suppose the walker begins at station 0 and tosses the coin a total of fifteen times. The first two tosses come up heads, and the third comes up tails. This sequence of three tosses is then repeated in exactly the same order for four additional cycles. The resulting deterministic trajectory is illustrated in Figure 4-9.

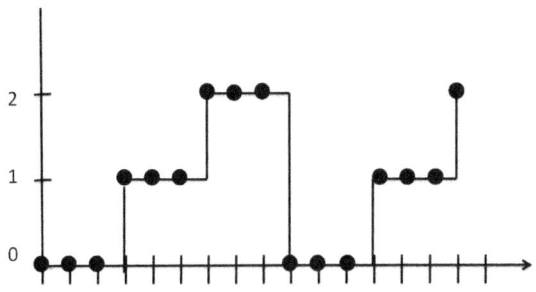

Figure 4-9. A trajectory for Figure 4-8

$$
\begin{pmatrix}
2/3 & 1/3 & 0 \\
0 & 2/3 & 1/3 \\
1/3 & 0 & 2/3
\end{pmatrix}
$$

Figure 4-10. Global transition matrix for the trajectory in Figure 4-9

The global transition matrix for this trajectory is presented in Figure 4-10. If the endpoints of the trajectory had been matched, the normalized eigenvector of this matrix would be comprised of the correct values of P(j). For the trajectory in Figure 4-9, the walker begins at station 0 and ends at station 2. Even though the endpoints are not matched, the normalized eigenvector of the matrix in Figure 4-10 can still be used to approximate the values of P(j). Solving the balance equations for this matrix yields the following result:

P(0) is approximately equal to $1/3 = .333$

P(1) is approximately equal to $1/3 = .333$

P(2) is approximately equal to $1/3 = .333$

Corollary 4.1 can be used to compute the exact values of P(j) for the trajectory in Figure 4-9 . Since the trajectory begins at station 0 and ends at station 2, $a = 0$, $b = 1$ and $c = 2$. Changing the final station from station 2 to station 0 creates an adjusted trajectory with matched endpoints. The global transition matrix for this adjusted trajectory is shown in Figure 4-11. Note that this matrix can be generated directly from the matrix in Figure 4-10 by applying equations (4-25) and (4-26) with V(1), the number of visits to station 1, set equal to six.

$$\begin{pmatrix} {}^2/_3 & {}^1/_3 & 0 \\ {}^1/_6 & {}^2/_3 & {}^1/_6 \\ {}^1/_3 & 0 & {}^2/_3 \end{pmatrix}$$

Figure 4-11. Global Transition Matrix for Adjusted Trajectory

Solving the balance equations for this matrix yields the following results:

P(0) is exactly equal to $2/5 = .400$

P(1) is exactly equal to $2/5 = .400$

P(2) is exactly equal to $1/5 = .200$

For this relatively short trajectory, the approximate solution obtained by assuming matched endpoints is rather poor. However, accuracy improves when the deterministic pattern depicted in Figure 4-9 is repeated multiple times. For example, if the length of the trajectory is increased from 16 to 97 (i.e., from 5 to 32 cycles of three tosses each), the initial station and final stations remain unchanged while the value of V(1) increases to 33. Changing the final station from station 2 to station 0 again creates an adjusted trajectory with matched endpoints. The global transition matrix for this longer adjusted trajectory is shown in Figure 4-12.

$$\begin{pmatrix} {}^2/_3 & {}^1/_3 & 0 \\ {}^1/_{33} & {}^2/_3 & {}^{10}/_{33} \\ {}^1/_3 & 0 & {}^2/_3 \end{pmatrix}$$

Figure 4-12. Global transition matrix for longer trajectory

Solving the balance equations for this matrix yields the following results:

P(0) is exactly equal to $33/96 = .343$

P(1) is exactly equal to $33/96 = .343$

P(2) is exactly equal to $30/96 = .313$

Note that the approximate solution obtained under the assumption of matched endpoints is now much closer to the actual solution shown above. As the length of the trajectory continues to increase, the approximate solution will converge to the exact solution.

This behavior is not limited to strictly periodic trajectories of the type illustrated in Figure 4-9. Convergence occurs for most models of practical interest, provided that the initial state of the trajectory is recurrent. Thus, for trajectories that are at least moderately long, it is reasonable for practitioners to use results derived under the assumption of matched endpoints, whether or not the endpoints of a trajectory are actually matched. This is similar in spirit to the use of steady state probability distributions to approximate the properties of stochastic processes that have been operating for long intervals of time but are not technically in steady state.

4.7.4 Matched endpoints and t-loops

The variables P(0), P(1), P(2) ... P(N) represent the proportion of visits the walker makes to each station over the course of a trajectory. The decision to exclude the final visit to the last station may seem somewhat arbitrary and a bit untidy. However, the decision clearly makes very little difference in most cases of practical interest. Moreover, in those rare cases where it is essential to account for the final visit to the last station, it is a trivial matter to adjust the values of P(j) provided that the identity of the last station (i.e., c) and the total length of the trajectory (i.e., V) are both known:

  simply replace $P(j)$ by: $P(j) \times V/(V+1)$ if $j \neq c$

  and replace $P(c)$ by: $P(c) \times V/(V+1) + 1/(V+1)$.

Another way to handle this issue is to reformulate the original problem so that the notion of initial and final stations is simply eliminated.   To motivate this alternative approach, recall that observational stochastics was originally characterized as the analysis of finite length trajectories with well defined endpoints. If the endpoints of a trajectory are matched, their actual identity is immaterial and does not enter into the analysis.

In such cases, it is possible to shift the focus of the analysis from conventional trajectories to trajectory loops (t-loops). To create a t-loop, begin with a trajectory with matched endpoints.   Then eliminate the final state and construct a circular trajectory in which the next to the last station of the original trajectory is followed by the first station of that trajectory.   As shown in Figure 4-13, this creates a new structure that is one station shorter than the original trajectory.

The order in which stations appear is the same in both the original trajectory and the associated t-loop.   However, a t-loop has no fixed beginning and no fixed end.   Instead, the stations simply reappear in the same order for an indefinite number of complete cycles. The starting point for these cycles is immaterial.

Note that the quantities $X(j,k)^{\#}$, $V(j)$, V, $X(j,k)$ and $P(j)$ all have exactly the same values in both the   t-loop and the original trajectory.   For the t-loop, there is no need to exclude the last station when evaluating these quantities since the last station is removed during the formation of the t-loop itself.

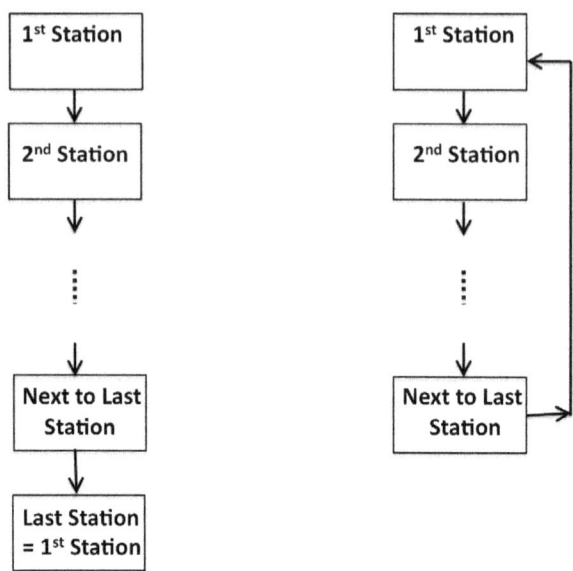

Figure 4-13. Constructing a t-loop

It should be clear that this shift of focus from trajectories with matched endpoints to the corresponding t-loops leaves every step of the analysis unchanged. The only special consideration in the case of t-loops is the implicit understanding that each analysis deals with one or more complete cycles around the entire loop. This corresponds to the understanding that each analysis of a conventional trajectory deals with one complete instance of that trajectory. In effect, the formal mathematical model upon which observational stochastics is based shifts from state transition diagrams, workloads and trajectories to state transition diagrams, workloads and t-loops.

T-loops are also interesting mathematical objects in their own right. Note that a t-loop is defined by specifying a sequence of states in some given order. The final state in this specification is assumed to be followed directly by the initial state to form a closed loop. Thus the direction in which time advances is implicit in the specification of a t-loop. If this direction is reversed, the values of $X(j,k)^{\#}$ and $X(j,k)$ will be altered; however, it is clear that the values of $P(j)$ will remain exactly the same. Note also that the new value of $X(j,k)^{\#}$ will be identical to the original value of $X(k,j)^{\#}$.

Two t-loops can be joined together (spliced) to form a longer t-loop. In the case of a *natural splice,* both loops must include a pair of successive states (e.g., a and b) in the same order. A natural splice of the two t-loops causes station a of the first loop to be followed by station b of the second loop. Similarly, station a of the second loop is followed by station b of the first loop to complete the splice.

When two t-loops are combined by a natural splice, the resulting values of $X(j,k)$ and $P(j)$ are easy to compute, provided the length of each loop is known. In the special case where the values of $X(j,k)$ and $P(j)$ are the same in each original loop, they will remain unchanged in the spliced loop.

It is also possible to define a *forced splice* constructed by joining two loops at arbitrary points. This complicates the procedure for determining the resulting values of $X(j,k)$. However, the resulting values of $P(j)$ can be determined using exactly the same procedure employed for natural splices. In particular, it is easy to see that the values of $P(j)$ remain unchanged whenever two t-loops with identical values of $P(j)$ are joined by a forced splice.

A single t-loop can also be split into two smaller t-loops via an unsplice operation. Little can be said regarding the outcome of an unsplice unless additional assumptions are introduced. T-loops may have other interesting mathematical properties that are as yet undiscovered.

4.8 Relationship to Markov chains

Readers familiar with the theory of Markov chains will notice a direct and obvious connection between the results derived in this chapter using observational stochastics and the corresponding results derived using classical Markovian assumptions. From a classical perspective, a time-homogeneous Markov chain or Markov model is a stochastic process (i.e., a sequence of random variables) whose most important properties are characterized by a transition matrix of the form shown in Figure 4-14.

$$
\begin{pmatrix}
\underline{X}(0,0) & \underline{X}(0,1) & \underline{X}(0,2) & \cdots & \underline{X}(0,N) \\
\underline{X}(1,0) & \underline{X}(1,1) & \underline{X}(1,2) & \cdots & \underline{X}(1,N) \\
\vdots & \vdots & \vdots & & \vdots \\
\vdots & \vdots & \vdots & & \vdots \\
\underline{X}(N,0) & \underline{X}(N,1) & \underline{X}(N,2) & \cdots & \underline{X}(N,N)
\end{pmatrix}
$$

Figure 4-14. Stepwise transition matrix: N+1 stations

In this example, each row in the matrix corresponds to a set of probabilities that relate the next random variable in the sequence to the current random variable. The probability that the next random variable will have the value k, given that the current random variable has the value j, is equal to $\underline{X}(j,k)$.

In effect, the values of $\underline{X}(j,k)$ make it possible to compute the probability distribution for each successive random variable in the sequence, given the probability distribution of the current random variable. If the initial distribution (at time zero) is also provided, it is possible to use the information in the stepwise transition matrix to generate the entire sequence of probability distributions in the Markov chain. Table 2-1 in Chapter 2 provides an example of these computations for a simple random walk.

If a Markov chain is in steady state, the probability distribution that characterizes the next random variable is identical to the probability distribution that characterizes the current random variable. Mathematically, this implies that the steady state probabilities $\underline{P}(0)$, $\underline{P}(1)$, $\underline{P}(2)$ and $\underline{P}(3)$ correspond to the normalized eigenvector of the stepwise transition matrix depicted in Figure 4-14. This is, of course, directly analogous to the way the values of P(j) are characterized in Theorem 4.1.

4.8.1 Stepwise and global transition matrices

There are, however, some fundamentally important differences between a classical (i.e., time-homogeneous) Markov chain and its observational counterpart. The most obvious difference is that the variables X(j,k) that make up the matrix in Figure 4-5 are defined as observable quantities of trajectories, while the variables $\underline{X}(j,k)$ that make up the matrix in Figure 4-14 represent probabilities that are not linked to directly observable quantities of any type.

Another important point is that the probabilities in Figure 4-14 regulate the step-by-step relationship between all successive pairs of random variables in the sequence that forms the Markov chain. These transition probabilities are required to be exactly the same for each step in the sequence. In contrast, the proportions that appear in Figure 4-5 are global values that apply to the trajectory as a whole. To highlight this difference, the term global transition matrix is used in conjunction with Figure 4-5 and the term stepwise transition matrix is used in conjunction with Figure 4-14.

4.8.2 A strictly periodic Markov chain

As will be discussed in Chapter 5, the step-by-step details that can be represented by traditional Markov chains are useful for analyzing certain aspects of system behavior. However, these extra details can sometimes create unwanted complications. The special random walk depicted in Figure 4-15 illustrates the nature of one such complication.

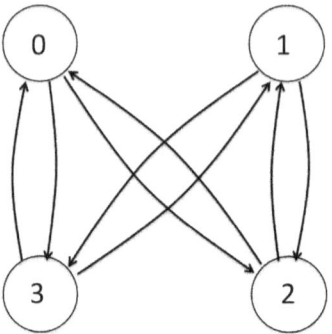

Figure 4-15. Periodic behavior

As usual, assume that the station-to-station transitions the walker makes are controlled by the toss of a fair coin. The special rules governing these transitions are as follows:

1. If the walker is currently at station 0, *heads* takes the walker to station 2 and *tails* takes the walker to station 3.

2. If the walker is currently at station 1, *heads* takes the walker to station 2 and *tails* takes the walker to station 3.

3. If the walker is currently at station 2, *heads* takes the walker to station 0 and *tails* takes the walker to station 1.

4. If the walker is currently at station 3, *heads* takes the walker to station 0 and *tails* takes the walker to station 1.

For the simple case where the probability of heads is always equal to one half, the stepwise transition matrix for this random walk is shown in Figure 4-16.

Suppose, as usual, the walker begins at station 2. After the first step, the walker is equally likely to be either at station 0 or at station 1. After the second step, the walker is equally likely to be at either station 2 or station 3. This cycle continues indefinitely. As a result of this strictly periodic behavior, the sequence of random variables that form this Markov chain never converges to a stable distribution as depicted in Table 2-1. Limiting values of $\underline{P}(0)$, $\underline{P}(1)$, $\underline{P}(2)$ and $\underline{P}(3)$ do not exist in this case.

$$
\begin{pmatrix}
0 & 0 & 1/2 & 1/2 \\
0 & 0 & 1/2 & 1/2 \\
1/2 & 1/2 & 0 & 0 \\
1/2 & 1/2 & 0 & 0
\end{pmatrix}
$$

Figure 4-16. Stepwise transition matrix for Figure 4-10

On the other hand, if Figure 4-16 is interpreted as a global transition matrix that has been obtained by observing an actual trajectory, the analysis presented in Section 4.6 is directly applicable. The values of P(0), P(1), P(2) and P(3) always exist and are given by the normalized eigenvector of the matrix in Figure 4-16. In this case, all four values of P(j) are simply equal to ¼.

The periodic nature of this particular random walk creates no problems whatsoever when the trajectory generated by the walker is analyzed using observational stochastics. By defining the values of X(j,k) and P(j) globally rather than on a step-by-step basis, observational stochastics is able to derive accurate results without having to deal with unnecessary complexities.

4.8.3 Physically unattainable initial conditions

It is important to remember that the stepwise transition matrix in Figure 4-16 still defines a valid Markov chain. This Markov chain does, in fact, have a steady state distribution. This distribution is [¼, ¼, ¼, ¼]. However, the sequence of probability distributions generated by the transition matrix in Figure 4-16 will never converge to [¼, ¼, ¼, ¼] if the walker starts at stations 0, 1, 2 or 3.

On the other hand, according to the standard mathematical definition of a Markov chain, the walker's position at each step of the walk is represented by a random variable. This includes the walker's starting position as well as all subsequent positions. Suppose the distribution that characterizes the walker's starting position is [¼, ¼, ¼, ¼]. In other words, consider a Markov chain characterized by Figure 4-16 that starts in steady state. This Markov chain will then remain in steady state forever since every probability distribution in the sequence will be identical to the first. This is directly analogous to the situation illustrated in Table 2-3 and Table 2-4 of Chapter 2.

While this reasoning is mathematically sound, it points out a fundamental difference between stochastic modeling and observational stochastics. In observational stochastics, there are four possible starting positions for the walker: stations 0, 1, 2 or 3. These are the only starting positions that are physically attainable. Moreover, every trajectory considered within observational stochastics represents a physically attainable sequence of stations.

In contrast, the assumption that the walker's starting position is characterized by the probability distribution [¼, ¼, ¼, ¼] is entirely legitimate within the context of a traditional Markov chain. This is true even though such a starting position cannot be incorporated into a conventional trajectory and thus corresponds to a physically unattainable assumption. This in no way affects the self-consistency and technical correctness of the underlying mathematics. However, it does call into question the connection between the resulting mathematical model and the real world phenomena being analyzed. The ability to obtain useful solutions without requiring the introduction of physically unattainable assumptions is yet another benefit of observational stochastics.

CHAPTER 5
Modeling Techniques and Examples

5.1 Construction and application of LCD models

Chapter 3 introduced the notion of LCD models through a simple example: a random walk in one dimension with reflecting barriers. This chapter presents a series of examples that involve the application of LCD models to systems whose behavior is somewhat more complex. The modeling techniques described here are quite general and extend well beyond these specific examples.

Explicit algebraic solutions are derived in most cases, but these are of secondary importance. The primary goal is to demonstrate how the basic building blocks that are available in observational stochastics can be used to solve a variety of problems.

5.1.1 Specifying the states of a model

The first step in the construction of any LCD model is to identify the set of states upon which the model will be based. For the random walks described in Chapter 3, state specification is both simple and intuitive: each station visited during a walk is represented by a separate state. Thus, there is a direct one-to-one relationship between states and physical locations.

State specification is not always so straightforward. Sections 5.2 through 5.6 of this chapter illustrate the way states can be used to represent logical rather than physical entities. These examples deal primarily with behavior that is driven by the number of successive visits the walker makes to the same location.

Section 5.7 presents an entirely different example that illustrates the distinction between good and poor choices in the specification of a model's states. In this example, states correspond directly to simple physical entities and locations. Nevertheless, a substantial degree of insight and creativity is still required to select the most efficient set of states for a model that is developed to solve an intriguing mathematical puzzle.

Note that the state specification process is essentially the same for both LCD models and traditional stochastic models. All the LCD models presented in this book have traditional stochastic counterparts that are based on identical sets of states.

5.1.2 Specifying the dynamic behavior of a model

Once a set of states has been identified, the next step in the development of an LCD model is to specify the dynamic behavior of the system being analyzed. In observational stochastics, the state of a system at any instant is a directly observable quantity. As a result, dynamic behavior can be characterized by trajectories that reflect the step-by-step changes in a system's observable state that take place over an interval of time.

Trajectories of this type can be specified by state transition diagrams or by finite state automata. For example, the state transition diagram shown in Figure 3-1 specifies the dynamic behavior of the simple random walk in Chapter 3. The finite state automaton specified by equations (3-44) through (3-51) provides an alternative format for specifying the same information.

In a traditional stochastic model, the state of a system at any instant is characterized by a probability distribution. As a result, a system's dynamic behavior can be specified by a sequence of probability distributions defined over an interval of time. For the traditional random walk in Chapter 2, step-by-step variations in these probability distributions are specified by equations (2-9) through (2-12). Equations of this type are used to characterize the dynamic behavior of all discrete time Markov processes.

5.1.3 Specifying the parameters of a model

Equations (3-44) through (3-51) and equations (2-9) through (2-12) are both used to specify dynamic behavior. However, these two sets of equations differ in one fundamentally important respect. The first set of equations is "parameter free" while the second incorporates the distributional parameter r as well as the implicit distributional assumption of independent Bernoulli trials (which determines the joint distribution of the random variables that characterize the sequence of turns the walker makes).

In contrast, equations (3-44) to (3-51) are expressed solely in terms of input symbols (⇑ and ⇓) and states (0, 1, 2 and 3). There are no numerical parameters and no implicit assumptions.

Since dynamic behavior is specified in a parameter free manner, the next step in the construction of an LCD model is to specify an appropriate set of parameters and modeling assumptions for the system being analyzed. One option is to employ no modeling assumptions at all (except for the technical assumption of matched endpoints). This leads to a set of maximally detailed parameters. For the simple random walk in Chapter 3, this maximally detailed set is: $R(0)$, $R(1)$, $R(2)$ and $R(3)$. Results such as equations (3-30) through (3-33) that are expressed in terms of these maximally detailed parameters are, in general, applicable to all trajectories that can be generated by the model's state transition diagram.

Solutions expressed in terms of maximally detailed parameters are of limited practical value because it is often unrealistic to assume that these detailed quantities can be estimated or predicted with an acceptable degree of accuracy. To address this legitimate concern, LCD models provide a mechanism for reducing the total number of required parameters. This reduction is achieved by introducing additional modeling assumptions that are expressed in terms of loose constraints.

Essentially, loose constraints are assumptions about relationships among the maximally detailed parameters of the original LCD model. For example, the assumption that each turn a walker makes is empirically independent of the walker's current location implies that all values of R(n) must be equal to a common value R. Solutions derived under this assumption depend only on the value of R and are valid for all trajectories that satisfy this particular loose constraint. The individual values of R(n) are immaterial.

The use of loose constraints to reduce the number of independent parameters in LCD models has no direct counterpart in traditional stochastic modeling. However, this process plays a crucially important role in observational stochastics.

Note that loose constraints must also be intuitively plausible. In other words, practitioners must have good reason to believe that the loose constraints incorporated into an LCD model are likely to be satisfied by trajectories that will be encountered during "what if" experiments that have not yet been conducted. This issue is discussed further in Sections 6.4 through 6.8. See especially the discussion of modeling risk in Section 6.7.1.

5.2 Counting successive visits to stations

The simple random walk analyzed in Chapter 3 provides the starting point for a series of increasingly complex examples presented in Sections 5.2 through 5.6. In these examples, states and transitions are added to the basic model to represent detailed aspects of the walker's behavior not considered previously.

Begin by examining Figure 5-1, which presents an alternative view of a simple random walk. Note that this diagram incorporates a special feature that is not present in Figure 3-1. The four physical stations visited during the walk are now represented by rectangles rather than circles. When the walker is at physical station j, the walker is also said to be in logical state j. Logical states, which are represented by circles, are one of the basic building blocks for both LCD models and traditional stochastic models. There is a one-to-one relationship between physical stations and logical states in Figure 5-1, but this relationship will change in some of the other examples presented in this chapter.

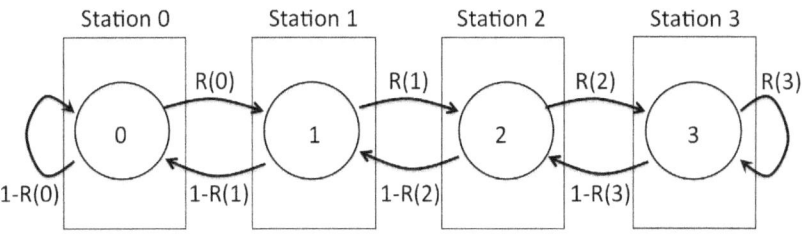

Figure 5-1. Labeling transition arrows with observable proportions

Another difference concerns the labels that are attached to the transition arrows in Figure 5-1. In Figure 3-1, these labels are input symbols that make it possible to interpret the original state transition diagram in Figure 2-1 as a finite state automaton. The labels on each arrow in Figure 5-1 serve a different purpose. They represent the proportion of time the walker selects the associated transition arrow when exiting from the originating state. In this particular diagram, R(j) represents the proportion of time the walker turns to the right when exiting from state j, and 1-R(j) represents the proportion of times the walker turns to the left.

The global transition matrix in Figure 5-2 provides a compact representation of the same information that appears in Figure 5-1. As discussed in Chapter 4, each row of this matrix is associated with a different state. The values in each column of a given row represent the proportion of time each possible destination is selected when the walker exits from the state associated with that row. These proportions correspond to the symbolic variables X(j,k) introduced in Chapter 4.

$$\begin{pmatrix} 1-R(0) & R(0) & 0 & 0 \\ 1-R(1) & 0 & R(1) & 0 \\ 0 & 1-R(2) & 0 & R(2) \\ 0 & 0 & 1-R(3) & R(3) \end{pmatrix}$$

Figure 5-2. Global transition matrix for Figure 5-1

As already noted, the initial objective when analyzing the random walk depicted in Figure 5-1 is to express the values of P(0), P(1), P(2) and P(3) as functions of the parameters R(0), R(1), R(2) and R(3). Equations (3-30) through (3-33) represent the desired solution. If all four values of R(j) are equal to R, the solution can be simplified substantially as shown in equations (3-26) through (3-29).

Although the values of P(j) are both useful and significant, they are not sufficient to answer all possible questions of interest. Problem 1 presents an example where these values are sufficient, and problem 2 presents a closely related example where they are not.

Problem 1 – Consider the proportion of turns that cause the walker to rebound off the reflecting barrier to the right of station 3. Express this proportion as a function of R.

Problem 2 – Consider the proportion of turns that are preceded by two or more successive rebounds off the reflecting barrier to the right of station 3. Express this proportion as a function of R.

To solve Problem 1, note that each rebound off station 3's reflecting barrier is generated by a visit to station 3 followed by a turn to the right. The proportion of visits to station 3 is P(3), and the proportion of those visits that are followed by a right turn is R(3). Thus $R(3) \times P(3)$ is the proportion of turns that cause the walker to rebound off the reflecting barrier to the right of station 3.

In Problem 1, the only information available to the analyst is R, the overall proportion of right turns. Thus, the solution to Problem 1 requires an assumption about the relationship between R and the individual values of R(j). If the direction of the next turn is empirically independent of the walker's current location, all four values of R(j) will be equal to R. This assumption implies that the solution to Problem 1 is given by $R \times P(3)$, where the value of $P(3)$ is specified by equation (3-29).

Thus the solution to Problem 1 is:

$$R \times P(3) = \frac{R \times [R/(1-R)]^3}{1 + [R/(1-R)] + [R/(1-R)]^2 + [R/(1-R)]^3} \qquad (5\text{-}1)$$

5.2.1 Successive encounters with the reflecting barrier

Determining the proportion of time the walker rebounds off station 3's reflecting barrier two or more times in succession is a more challenging problem. To solve Problem 2, it is necessary to create a general method for counting the number of successive right turns the walker makes after his or her initial transition from station 2 to station 3.

These successive right turns can be counted by adding special states to the original state transition diagram. As shown in Figure 5-3, the idea is to replace state 3 in the original state transition diagram by three special states: state 30, state 31 and state 32. These states, which are all associated with physical station 3, enable the model to keep track of the number of successive rebounds.

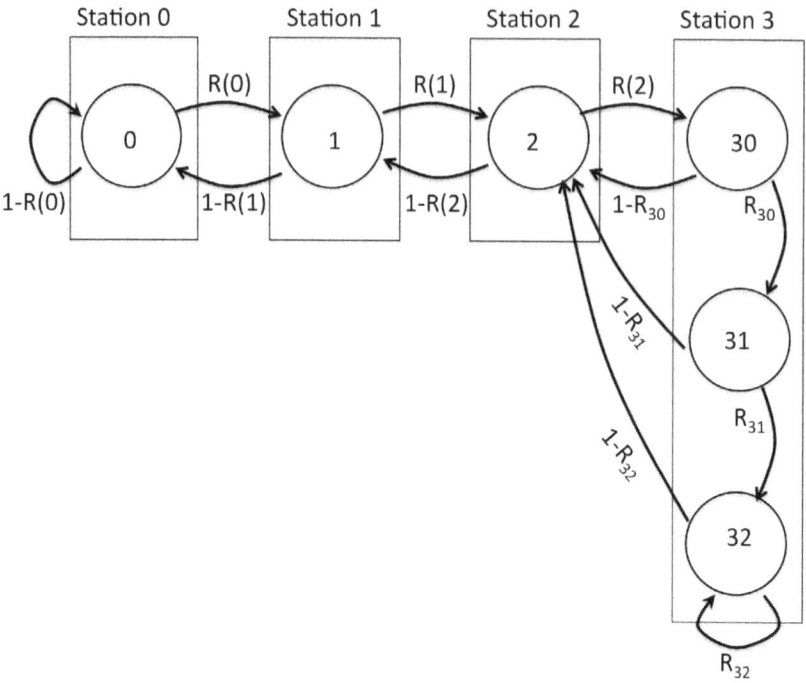

Figure 5-3. Adding states to count successive rebounds from Station 3

Note that an initial transition from station 2 to station 3 places the walker in state 30. The walker then has two options: turn left and return to state 2, or turn right, rebound off the reflecting barrier, and enter state 31. Thus, to reach state 31, the walker must rebound once off the reflecting barrier.

As shown in Figure 5-3, state 32 is reached from state 31 by a second successive rebound off the reflecting barrier. Once in state 32, additional right turns cause the walker to continue rebounding off the reflecting barrier and returning to state 32. Thus, the total number of times the walker rebounds off the reflecting barrier two or more times in succession is simply the total number of visits the walker makes to state 32. Left turns from states 30, 31 and 32 always return the walker to station 2 and effectively reset the mechanism that counts the number of successive right turns.

The next step in the analysis is to write down the balance equations associated with the model depicted in Figure 5-3. Some additional information is needed before this can be done. Recall that the solution to Problem 1 employed the assumption that the proportion of times the walker turns right is the same for all four physical stations, and is represented by the parameter R: in other words, $R(0) = R(1) = R(2) = R(3) = R$.

This simplifying assumption applies to physical station 3 as a whole, but does not necessarily apply to logical states 30, 31 and 32 within physical station 3. For example, it is possible that right turns are relatively less likely from state 32 since the walker has already made three right turns in a row before reaching this state. To allow for this possibility, a separate parameter can be used to specify the proportion of times the walker turns to the right after leaving each special state. Three parameters are required: R_{30}, R_{31} and R_{32}. The global transition matrix for this model will then have the form shown in Figure 5-4.

$$\begin{pmatrix} 1-R & R & 0 & 0 & 0 & 0 \\ 1-R & 0 & R & 0 & 0 & 0 \\ 0 & 1-R & 0 & R & 0 & 0 \\ 0 & 0 & 1-R_{30} & 0 & R_{30} & 0 \\ 0 & 0 & 1-R_{31} & 0 & 0 & R_{31} \\ 0 & 0 & 1-R_{32} & 0 & 0 & R_{32} \end{pmatrix}$$

Figure 5-4. Simplified global transition matrix for Figure 5-2

5.2.2 Algebraic solution

The algebraic solutions derived in this section are of secondary importance. Some readers may prefer to skip this material and proceed directly to Section 5.2.3.

Equations (5-2) through (5-7) are the balance equations associated with Figures 5-3 and 5-4. These balance equations can be derived directly from Figure 5-3 by examining the possible ways the walker can arrive at each of the six states in the model. These possibilities are as follows:

There are two ways the walker can arrive at state 0: start from state 0, turn left, and rebound off the reflecting barrier; or start from state 1 and turn left. These two options lead to equation (5-2).

$$P(0) = (1-R) \times P(0) + (1-R) \times P(1) \tag{5-2}$$

Similarly, there are two ways the walker can arrive at state 1: start from state 0 and turn right; or start from state 2 and turn left. These two options lead to equation (5-3).

$$P(1) = R \times P(0) + (1-R) \times P(2) \tag{5-3}$$

In the case of state 2, there are four ways the walker can arrive: start from state 1 and turn right; or start from states 30, 31 or 32 and turn left. These four options lead to equation (5-4).

$$P(2) = RP(1) + (1-R_{30})P(30) + (1-R_{31})P(31) + (1-R_{32})P(32)$$

$$\tag{5-4}$$

State 30 is simpler since only one possibility exists: start at state 2 and turn right. This leads to equation (5-5).

$$P(30) = R \times P(2) \tag{5-5}$$

State 31 is similar to state 30: the only possibility is to start at state 30, turn right, and rebound off the reflecting barrier. This leads to equation (5-6).

$$P(31) = R_{30} \times P(30) \tag{5-6}$$

Finally the walker has two ways of reaching state 32: start at state 31 turn right, and rebound off the reflecting barrier; or start at state 32 turn right, and rebound off the reflecting barrier.

$$P(32) = R_{31} \times P(31) + R_{32} \times P(32) \tag{5-7}$$

Equations (5-2) through (5-7), along with the familiar normalization constraint, yield the solution shown in equations (5-8) through (5-13).

Let $Q = R / (1 - R)$

$$P(0) = \frac{1}{1 + Q + Q^2 + RQ^2 + R_{30}RQ^2 + [R_{31} / (1 - R_{32})]R_{30}RQ^2}$$

$$\tag{5-8}$$

$$P(1) = Q \times P(0) \tag{5-9}$$

$$P(2) = Q^2 \times P(0) \tag{5-10}$$

$$P(30) = R \times Q^2 \times P(0) \tag{5-11}$$

$$P(31) = R_{30} \times R \times Q^2 \times P(0) \tag{5-12}$$

$$P(32) = R_{30} \times R \times Q^2 \times [R_{31} / (1 - R_{32})] \times P(0) \tag{5-13}$$

As noted previously, each visit to state 32 must be preceded by two or more successive rebounds off the reflecting barrier to the right of station 3. Since P(32) is the proportion of visits the walker makes to state 32 over the course of the entire trajectory, P(32) must also be the solution to Problem 2.

However, the expression for P(32) given by equation (5-13) is a function of R_{30}, R_{31}, R_{32} and R. The solution to Problem 2 must be expressed entirely as a function of R, the overall proportion of right turns. Thus, to complete the analysis, it is also necessary to assume that R_{30}, R_{31} and R_{32} are all equal to R. This is, of course, yet another example of empirical independence. In combination with the assumption that R(0), R(1) and R(2) are all equal to R, this assumption implies that equations (5-8) through (5-13) can be simplified as follows:

$$P(0) = \frac{(1-R)^3}{1-2R+2R^2} \tag{5-14}$$

$$P(1) = \frac{R \times (1-R)^2}{1-2R+2R^2} \tag{5-15}$$

$$P(2) = \frac{R^2 \times (1-R)}{1-2R+2R^2} \tag{5-16}$$

$$P(30) = \frac{R^3 \times (1-R)}{1-2R+2R^2} \tag{5-17}$$

$$P(31) = \frac{R^4 \times (1-R)}{1-2R+2R^2} \tag{5-18}$$

$$P(32) = \frac{R^5}{1-2R+2R^2} \tag{5-19}$$

Thus, provided that R(0), R(1) and R(2) are equal to R, and that R_{30}, R_{31} and R_{32} are also equal to R, the solution to Problem 2 is the value of P(32) given by equation (5-19).

5.2.3 Level of detail versus risk

Even though the models in Figures 5-1 and 5-3 represent the same physical system, the risks associated with predictions derived from these two models are not identical. Predictions based on equation (5-19), which is derived from the model in Figure 5-3, are riskier than predictions based on equation (5-1).

To understand why, consider any trajectory that satisfies the assumptions required to derive equation (5-19). The values of $R(0)$, $R(1)$ and $R(2)$ for such a trajectory must all be equal to R. In addition, the values of R_{30}, R_{31} and R_{32} must also be equal to R. These assumptions imply that the value of $R(3)$ must be equal to R as well. Thus, if a trajectory satisfies the assumptions required to derive equation (5-19), it must also satisfy the assumptions required to derive equation (5-1).

The converse is not true. Consider a trajectory that satisfies the assumptions required to derive equation (5-1). The values of $R(0)$, $R(1)$, $R(2)$ and $R(3)$ for this trajectory must all be equal to R. This imposes certain constraints on the values of R_{30}, R_{31} and R_{32}, but does not ensure that each of these proportions is also equal to R. Thus the solution to Problem 2 expressed by equation (5-19) may or may not be valid for this trajectory.

Intuitively speaking, the region in which equation (5-19) provides the correct solution to Problem 2 is smaller than – and contained entirely within – the region in which equation (5-1) provides the correct solution to Problem 1. This means that the chance of equation (5-19) being incorrect is greater than the chance of equation (5-1) being incorrect. In practical terms, this implies that the solution to Problem 2 given by equation (5-19) is riskier than the solution to Problem 1 given by equation (5-1).

This form of risk does not arise when working with a standard Markov model of a random walk because the same model can be used to solve both Problem 1 and Problem 2. The stepwise transition matrix for this model has the structure shown in Figure 5-2. As in the observational case, it is necessary to assume that $R(0)$, $R(1)$, $R(2)$ and $R(3)$ are all equal to R. This is sufficient to derive the stochastic counterpart of equation (5-1), which solves Problem 1.

This Markov model includes the assumption that, for each step the walker takes, the probability of turning right is always equal to R. Thus the probability that the walker is at station 3 and that the next n turns are to the right (causing n successive rebounds off the reflecting barrier) is simply $R^n \times \underline{P}(3)$. Setting n equal to 1 produces the solution to Problem 1, and setting n equal to 2 produces the solution to Problem 2.

Since this same argument can be employed with any value of n, the assumptions that underlie the traditional Markov model of a random walk are actually quite powerful. However, this power obscures the point that predictions regarding the probability of n or more successive rebounds off station 3's reflecting barrier become inherently riskier as n increases. In contrast, the additional assumptions and risks associated with each increase in the value of n are made explicit in observational stochastics.

Similar comments are applicable to many other areas of applied probability, where predictions of rare events are sometimes derived through technically correct analyses of models that are based on deceptively powerful mathematical assumptions. The success of such models when making low risk predictions does not necessarily ensure that these same models will be equally successful when predicting rare events (e.g., the black swan events discussed by Taleb (2007)). Once again, this increased level of

risk is difficult to discern when employing traditional probabilistic models.

One additional point regarding the model in Figure 5-3 is also
relevant. As noted in Section 1.6.4, observational stochastics
enables analysts to carry out sensitivity analyses to assess the
magnitude of the error introduced when observational assumptions
are violated.

In the case of Problem 2, it is possible to obtain an exact solution
when the values of R_{30}, R_{31} and R_{32} are not all equal to R. This
exact solution can be used to examine the impact of assuming that
$R_{30} = R_{31} = R_{32} = R$ in cases where this assumption is only valid to
$\pm 5\%$. Analyses of this type can be used to strengthen the degree
of confidence in the original solution.

5.3 Limits on successive visits

Figures 5-1 and 5-3 illustrate the way special states can be used to
increase the level of detail in a model. The actual behavior of the
walker is exactly the same in both figures, but additional
information regarding the walker's behavior is represented in
Figure 5-3.

A similar approach can be used to represent examples where the
walker's behavior is slightly different in each case. For example,
suppose there is a fixed limit to the number of times the walker
will rebound successively off station 3's reflecting barrier. In
particular, suppose that after two successive rebounds the walker
decides to stop turning right and instead proceeds to station 2 on
the next turn. The state transition diagram in Figure 5-5
represents the walker's behavior in this case. The solution to the
balance equations for this model is presented in equations (5-20)
through (5-25). As one would expect, the value of P(32) is smaller
in this case, causing other values of P(k) to be relatively larger.

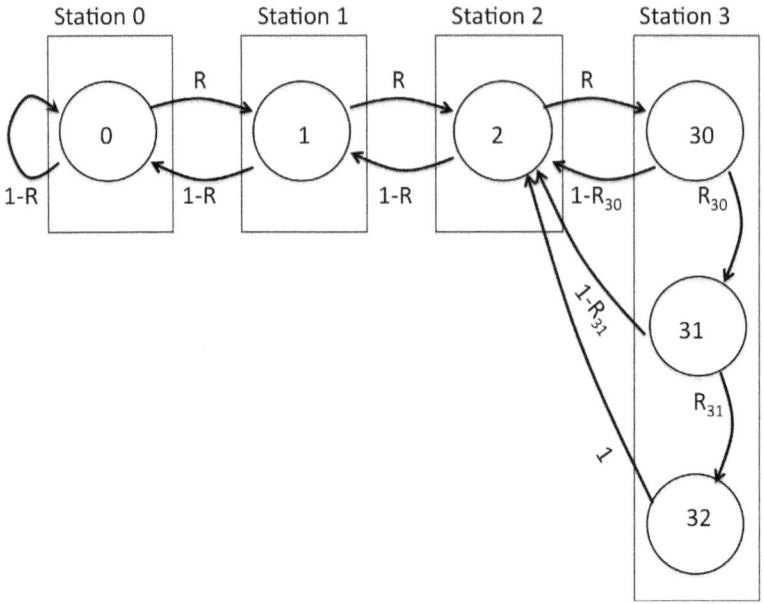

Figure 5-5. Maximum of two successive rebounds

Let $Q = R / (1 - R)$

$$P(0) = \frac{1}{1 + Q + Q^2 + RQ^2 + R_{30}RQ^2 + R_{31}R_{30}RQ^2} \qquad (5\text{-}20)$$

$$P(1) = Q \times P(0) \qquad (5\text{-}21)$$

$$P(2) = Q^2 \times P(0) \qquad (5\text{-}22)$$

$$P(30) = R \times Q^2 \times P(0) \qquad (5\text{-}23)$$

$$P(31) = R \times Q^2 \times P(0) \qquad (5\text{-}24)$$

$$P(32) = R_{31} \times R_{30} \times R \times Q^2 \times P(0) \qquad (5\text{-}25)$$

Figure 5-6 represents another modification of the walker's behavior. In this case, the walker always rebounds off the reflecting barrier at least once after entering station 3. The two variations represented by Figures 5-5 and 5-6 correspond to upper and lower bounds on the number of successive rebounds off station 3's reflecting barrier. The solution for this case is specified by equations (5-26) through (5-31).

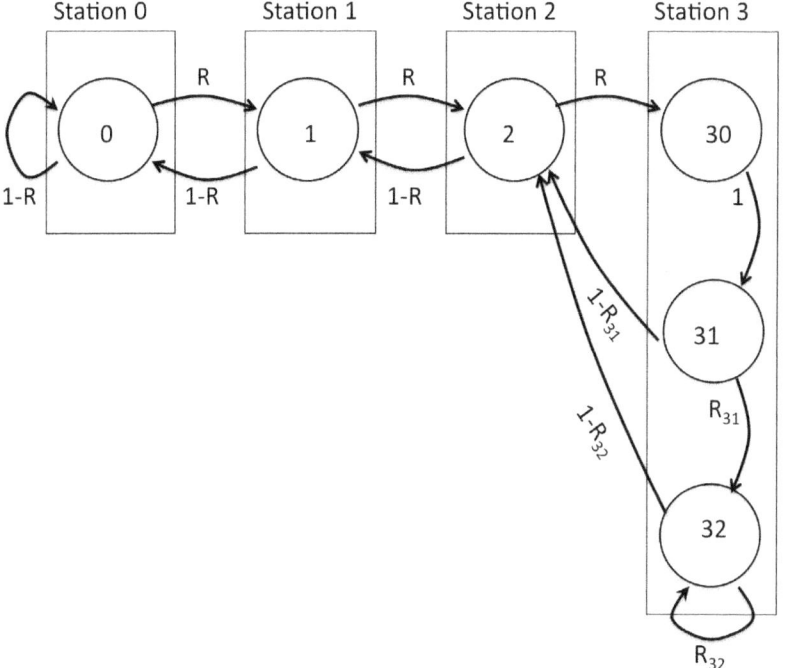

Figure 5-6. Minimum of one rebound

Let $Q = R / (1 - R)$

$$P(0) = \frac{1}{1 + Q + Q^2 + 2R \times Q^2 + R \times \left[R_{31} / (1 - R_{32}) \right] \times Q^2} \qquad (5\text{-}26)$$

$$P(1) = Q \times P(0) \qquad (5\text{-}27)$$

$$P(2) = Q^2 \times P(0) \qquad (5\text{-}28)$$

$$P(30) = R \times Q^2 \times P(0) \qquad (5\text{-}29)$$

$$P(31) = R \times Q^2 \times P(0) \qquad (5\text{-}30)$$

$$P(32) = \left[\frac{R_{31} \times R}{1 - R_{32}} \right] \times Q^2 \times P(0) \qquad (5\text{-}31)$$

5.4 Multiple states in series: method of stages

The bounds in Figures 5-5 and 5-6 can, of course, be combined into a single model as shown in Figure 5-7. In this particular example, a walker entering station 3 always goes from state 30 to state 31 to state 32 before exiting. These three states can be regarded as stages that the walker passes through while at station 3.

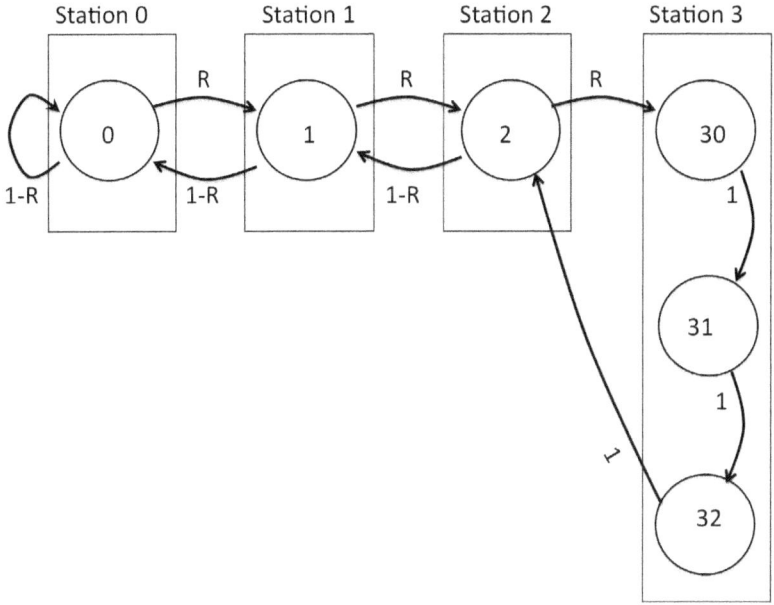

Figure 5-7. Always exactly two rebounds (three stages)

A more general example of stages appears in Figure 5-8. In this case, a walker entering station 3 goes through a series of k+1 stages. At each stage, the walker can continue on to the next (which corresponds to a right turn) or exit from station 3 and return to station 2 (which corresponds to a left turn). If the walker proceeds through all k+1 stages, the next transition always takes the walker back to station 2. There is no possibility of returning a second time to state 3k by rebounding off a reflecting barrier.

In this very general setting, it is possible to use the parameters R_{30}, R_{31}, R_{32} ... R_{3k-1} to represent a wide variety of patterns: for example, cases where most departures occur during the initial, middle or final stages, or during some combination of initial and final stages with few departures in the middle. The continuous time counterparts of these mechanisms are known as phase-type distributions (Stewart 2009, p. 155). Erlang and Coxian distributions (Erlang 1917), (Cox 1955), which are both examples of phase-type distributions, are closely linked to the general structure of states within station 3 of Figure 5-8.

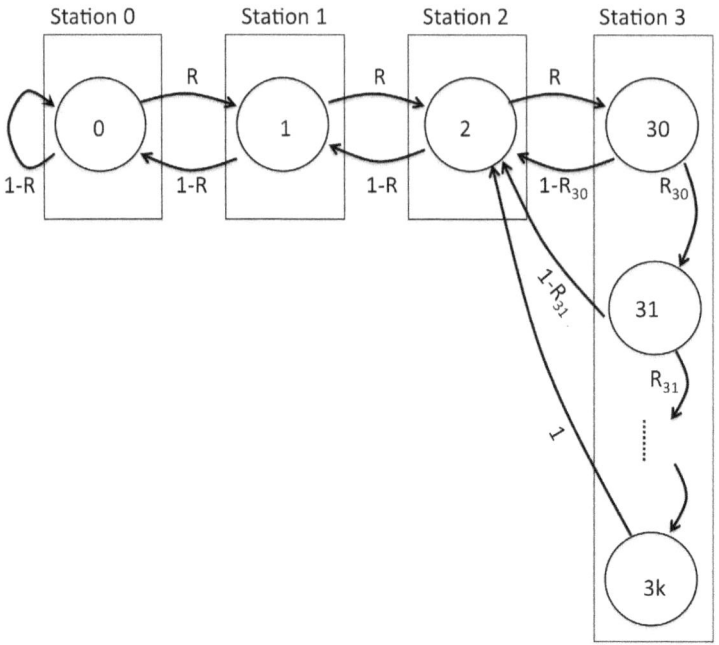

Figure 5-8. Maximum of k-1 rebounds

5.5 Two states in parallel: hyper-exponential mixtures

In Figure 5-8 the walker passes through a sequence of states in series before exiting from station 3. Another useful option is to permit the walker to proceed directly to any member of a set of states that are arrayed in parallel. Figure 5-9 illustrates this option for a simple case involving only two states: state 30 and state 31. As in the case of Figure 5-2, the parameters R_{30} and R_{31} represent the proportion of time the walker turns right and rebounds off the reflecting barrier after exiting from these states. The new parameter q represents the proportion of time the walker proceeds directly to state 30 when entering station 3. This implies that $1-q$

is the proportion of time the walker proceeds directly to state 31 when entering station 3.

Structures of this type are useful for representing certain special types of behavior. For example, suppose that most entrances to station 3 are followed almost immediately by a return to station 2 after a very small number of rebounds off the reflecting barrier. However, on rare occasions, suppose that the walker enters station 3 and rebounds off the reflecting barrier a very large number of times before returning to station 2.

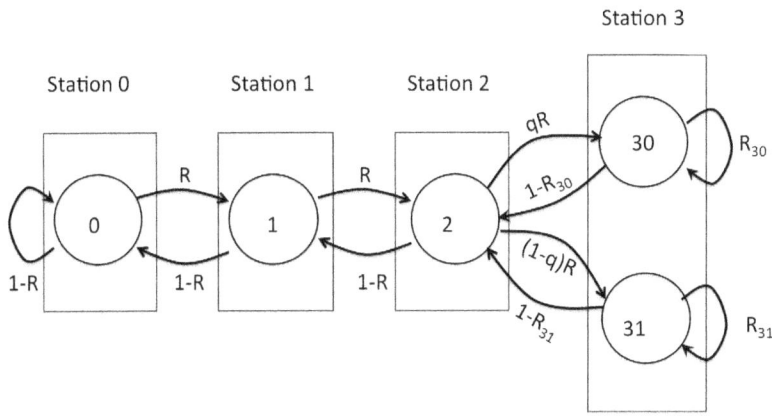

Figure 5-9. Two stages in parallel

This bi-modal behavior can be represented by setting the value of q very close to 1 (so that the walker almost always enters state 30 after exiting from station 2 and turning right) and by setting R_{30} close to zero (so that the walker is likely to return quickly to station 2 in this case). In addition, set the value of R_{31} close to 1 so that, on those rare occasions when the walker does enter state 31, there are likely to be a large number of rebounds before the walker finally returns to station 2.

Continuous time counterparts of this simple mechanism can be used to represent a number of real world phenomena that are sometimes characterized as being heavy-tailed or hyper-exponential. The hyper-exponential distribution, which is a mixture (i.e., a weighted average) of two exponential distributions, is another important example of a phase-type distribution (Stewart 2009).

5.6 Remembering past history in memoryless systems

Markov chains are often characterized as being memoryless because the rules that govern step-by-step transitions depend only on the current state and the probabilities in the stepwise transition matrix. Prior states have no impact on these rules. Since only the present matters, the past is effectively forgotten and the informal notion of being "memoryless" is appropriate.

The finite state automata used in observational stochastics share this memoryless characteristic. The next state in a trajectory depends only on the current state and the next input symbol in the workload. Prior states are once again immaterial. The rules that govern state-to-state transitions in observational stochastics are deterministic rather than probabilistic, but the rules are equally memoryless in both cases.

Despite their memoryless nature, there is a simple way to include information about past history in both types of models. Essentially, the idea is to add special states that can be used to encode historical information. The model depicted in Figure 5-3 provides an example of this technique: if the walker is currently in state 32, the walker's immediate history must include two or more successive encounters with station 3's reflecting barrier. In the simpler model shown in Figure 5-1, knowing that the walker is currently in station 3 provides no information about the walker's past encounters with the reflecting barrier. Both models are memoryless, but the special states that have been added to the model in Figure 5-3 provide information about past history.

The random walk depicted in Figure 5-10 exemplifies a different type of past history. In this example, there are only three stations in the walk: station 0, station 1 and station 2. The walker's behavior at station 0 is exactly the same as in the previous examples: a right turn after exiting from station 0 takes the walker to station 1, and a left turn causes the walker to rebound off the reflecting barrier and return to station 0.

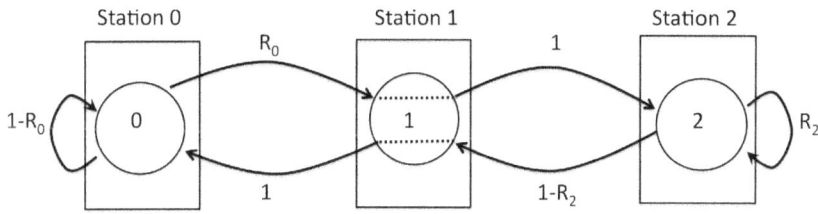

Figure 5-10. Behavior that is dependent on a prior state

Station 1 operates in a special manner. If the walker arrives at station 1 from station 0, the walker always turns right and proceeds to station 2. This new type of behavior is indicated by a dashed line connecting the two transitions. On the other hand, if the walker arrives at station 1 from station 2, the walker always turns left and proceeds to station 0. Once again, a dashed lined connecting the two transitions represents this behavior.

Station 2 operates in the same manner as station 3 in the prior examples: a left turn after exiting from station 2 takes the walker to station 1, and a right turn causes the walker to rebound off the reflecting barrier and return to station 2.

Despite the new behavior pattern at station 1, it is still possible to use observational stochastics to analyze the trajectories associated with the simple three state model in Figure 5-10. The global transition matrix for this model is shown in Figure 5-11. Note that the numbers of left and right exits from station 1 are always the same for a trajectory with matched endpoints. Thus the two non-zero values in the middle row of this matrix are always equal to ½.

$$\begin{pmatrix} 1-R_0 & R_0 & 0 \\ \frac{1}{2} & 0 & \frac{1}{2} \\ 0 & 1-R_2 & R_2 \end{pmatrix}$$

Figure 5-11. Global transition matrix

The balance equations for this model are:

$$P(0) = (1-R_0) \times P(0) + \frac{1}{2} P(1) \tag{5-32}$$

$$P(1) = R_0 \times P(0) + (1-R_2) \times P(2) \tag{5-33}$$

$$P(2) = \frac{1}{2} P(1) + R_2 \times P(2) \tag{5-34}$$

Equations (5-32) through (5-34), along with the standard normalization constraint, yield the following solution.

$$P(0) = \frac{1-R_2}{1-R_2 + 3R_0 - 2R_0 \times R_2} \tag{5-35}$$

$$P(1) = \frac{2R_0 \times (1-R_2)}{1-R_2 + 3R_0 - 2R_0 \times R_2} \tag{5-36}$$

$$P(2) = \frac{R_0}{1-R_2 + 3R_0 - 2R_0 \times R_2} \tag{5-37}$$

When analyzed within the framework of observational stochastics, the special behavior of the walker at station 1 is entirely immaterial. The expressions for P(0), P(1) and P(2) in equations (5-35) through (5-37) are guaranteed to be exactly correct for any trajectory generated by a walker with this pattern of behavior, provided the endpoints of the trajectory are matched.

On the other hand, the walker's behavior at station 1 makes it inappropriate to use Figure 5-10 as the basis for a Markov model. To see why, suppose that the matrix in Figure 5-11 is reinterpreted as a stepwise transition matrix. This implies that each time the walker enters station 1, the probability of turning right or left is always equal to ½. In a valid Markov model, this must be true regardless of the walker's past history. However, this assumption is clearly not valid for the model in Figure 5-10 since the direction of the walker's next turn depends entirely on whether the walker arrived at station 1 from station 0 or from station 2.

Additional states can, of course, be added to the model in Figure 5-10 to keep track of the walker's location just before arriving at station 1. The two possibilities can be represented by two special states: state 1.0 for arrivals from station 0, and state 1.2 for arrivals from station 2. Figure 5-12 illustrates an extended state transition diagram that incorporates these special states and can be used as the basis for a valid Markov model of this random walk.

Figure 5-13 displays the stepwise transition matrix for this model. Equations (5-38) through (5-41) present the corresponding balance equations.

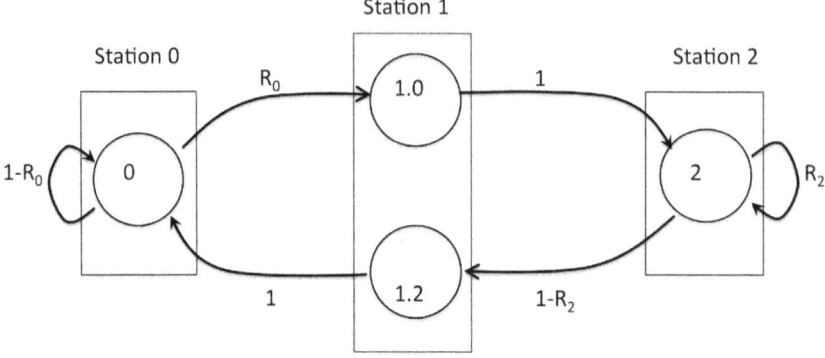

Figure 5-12. Extended model with memoryless behavior

$$\begin{pmatrix} 1-R_0 & R_0 & 0 & 0 \\ 0 & 0 & 0 & 1 \\ 1 & 0 & 0 & 0 \\ 0 & 0 & 1-R_2 & R_2 \end{pmatrix}$$

Figure 5-13. Stepwise transition matrix

$$P(0) = (1 - R_0) \times P(0) + P(1.2) \tag{5-38}$$

$$P(1.0) = R_0 \times P(0) \tag{5-39}$$

$$P(1.2) = (1 - R_2) \times P(2) \tag{5-40}$$

$$P(2) = P(1.0) + R_2 \times P(2) \tag{5-41}$$

The solution is:

$$P(0) = \frac{1 - R_2}{1 - R_2 + 3R_0 - 2R_0 \times R_2} \qquad (5\text{-}42)$$

$$P(1.0) = \frac{R_0 \times (1 - R_2)}{1 - R_2 + 3R_0 - 2R_0 \times R_2} \qquad (5\text{-}43)$$

$$P(1.2) = \frac{R_0 \times (1 - R_2)}{1 - R_2 + 3R_0 - 2R_0 \times R_2} \qquad (5\text{-}44)$$

$$P(2) = \frac{R_0}{1 - R_2 + 3R_0 - 2R_0 \times R_2} \qquad (5\text{-}45)$$

Equations (5-42) through (5-45) are the result of a valid Markov model. On the other hand, equations (5-35) through (5-37) are the result of a three-state model that does not satisfy traditional Markovian assumptions because it is not memoryless. Nevertheless, the value of P(0) in equation (5-42) is identical to the value of P(0) in equation (5-35). The same is true for the values of P(2) in equations (5-45) and (5-35). Finally, the value of P(1) in equation (5-36) is equal to the sum of P(1.0) and P(1.2) from equations (5-43) and (5-44).

Thus, even though the extra detail incorporated into Figure 5-12 is necessary to create a valid Markov model, this additional detail has no effect on the accuracy of the resulting predictions. An invalid Markov model based on Figure 5-10 and the matrix in Figure 5-11 still predicts P(0), P(1) and P(2) correctly. This provides yet another example of how a traditional stochastic model can generate accurate predictions when the assumptions required by the model are violated.

5.7 Structuring a model: the umbrella problem

The examples in Sections 5.2 through 5.6 are all variations of the same basic technique, which consists of adding special states to a model so that additional details can be represented. This section deals with a different modeling skill: identifying the most appropriate set of states to use when structuring a model.

The material in this section is adapted from an example described by Bertsekas and Tsitsiklis (2002, p. 329). The example involves a professor who walks from home to school each day over the course of a semester.

The professor is concerned about getting wet during his daily commute, so he begins the semester by placing one umbrella at his home and one in his office. This guarantees that the professor will always have an umbrella available on the first rainy day. However, the professor is absent minded and never remembers to bring an umbrella back to its original location on sunny days. Thus, the professor may get wet on subsequent trips if he happens to be caught with no umbrella on a rainy day.

Assume that R represents the proportion of trips that begin while it is raining. The problem is to express W, the proportion of times the professor actually gets wet, as a function of R. To simplify the analysis, assume that it never begins to rain midway through a trip.

The fact that the professor travels back and forth between home and school is reminiscent of the random walks considered previously. However, the proportion of visits that the professor makes to each physical location is not the main issue here. Instead, the primary objective of the umbrella problem is to determine the fraction of trips that begin when it is raining and no umbrellas are available. The purpose of building a model is to evaluate the proportion of trips that begin under these circumstances.

To construct any model, it is necessary to begin by identifying a set of states and their associated transitions. In the case of the umbrella problem, two different models can be formulated: the first is straightforward, but somewhat cumbersome and overly complex. The second, described in Bertsekas and Tsitsiklis (2002), is elegant and insightful.

For the first model, simply define each state as a combination of the professor's location and the number of available umbrellas. Since there are two possible locations (H=home and S=school) and three possibilities for the number of available umbrellas (0, 1 or 2), a total of six states are required: H0, H1, H2, S0, S1, S2.

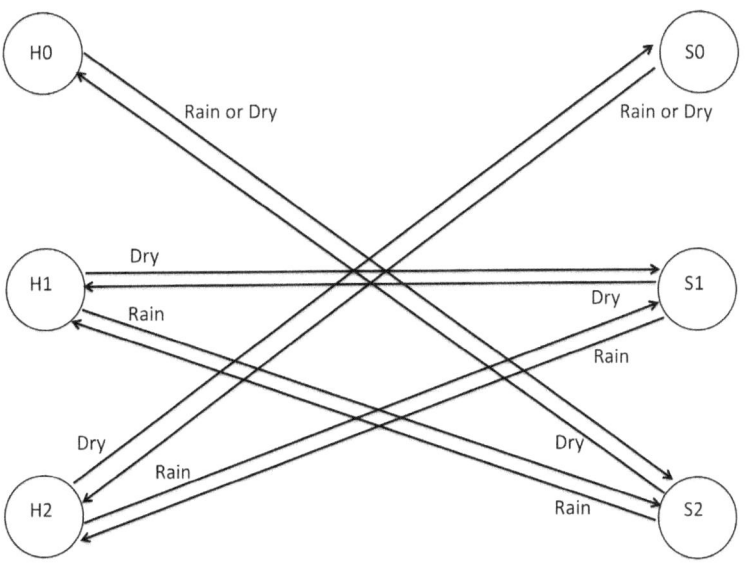

Figure 5-14. Umbrella problem - complex model

The state-to-state transitions are determined by whether it is raining (Rain) or dry (Dry) at the start of each trip. These factors can be regarded as input symbols for a finite state automaton. This representation leads to the state transition diagram shown in Figure 5-14.

Although this diagram is perfectly valid, it is more elaborate than necessary for the problem at hand. Determining the level of detail to include in a model is a critically important skill for practitioners involved in the analysis of real systems. This skill is entirely distinct from the mathematical proficiency required to solve a model (i.e., to derive algebraic expressions for the values of P(j)) once the model has been formulated).

The key to developing a simpler alternative model for the umbrella problem is recognizing that the objective is not to determine the proportion of time the professor visits each of the six states in Figure 5-14. Instead, the objective is to determine the proportion of time the professor gets wet. In order to get wet, two conditions must hold at the start of a trip: the number of available umbrellas must be equal to zero, and it must be raining. The actual location of the professor at such a time (home or school) is completely immaterial.

To generate the simplified model described by Bertsekas and Tsitsiklis, focus exclusively on the number of umbrellas available to the professor at the start of each trip. This number is either 0, 1 or 2. These three possibilities correspond to the three states of the simplified model. The state-to-state transitions for this model are shown in Figure 5-15.

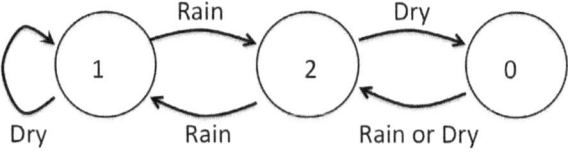

Figure 5-15. Umbrella problem - simple model

The next step in the solution process is to write down the global transition matrix that is associated with Figure 5-15. Recall that R is the overall proportion of rainy trips during the entire semester. To complete the global transition matrix, it is necessary to specify two related quantities: R_1, the proportion of rainy trips that begin when the professor is in state 1 (with one available umbrella); and R_2, the proportion of rainy trips that begin when the professor is in state 2 (with two available umbrellas). Note that R_0 does not appear in the global transition matrix because transitions out of state 0 always lead to state 2, whether it is raining or not.

$$\begin{pmatrix} 0 & 0 & 1 \\ 0 & 1-R_1 & R_1 \\ 1-R_2 & R_2 & 0 \end{pmatrix}$$

Figure 5-16. Global transition matrix: general case

Generally speaking, the number of available umbrellas (i.e., the current state) should not have an influence on the weather conditions at the start of a trip. Thus it is reasonable to assume that $R = R_0 = R_1 = R_2$. This is, of course, equivalent to assuming that the observed proportion of rainy trips is empirically independent of the number of available umbrellas. The balance equations for this case are equations (5-46) through (5-48). The solution is given by equations (5-49) through (5-51).

$$P(0) = (1-R) \times P(2) \tag{5-46}$$

$$P(1) = (1-R) \times P(1) + R \times P(2) \tag{5-47}$$

$$P(2) = R \times P(1) + P(0) \tag{5-48}$$

$$P(0) = \frac{1-R}{3-R} \tag{5-49}$$

$$P(1) = \frac{1}{3 - R} \qquad (5\text{-}50)$$

$$P(2) = \frac{1}{3 - R} \qquad (5\text{-}51)$$

In order for the professor to get wet, he must start a trip from state 0 when it is raining. Equation (5-49) states that the overall proportion of trips that start from state 0 is equal to (1-R)/(3-R). The proportion of those trips that begin while it is raining is R_0. Under the assumption of empirical independence, R_0 is equal to R. Thus the proportion of times the professor gets wet is:

$$W = R \times \frac{1 - R}{3 - R} \qquad (5\text{-}52)$$

This equation passes the test of reasonableness. If it never rains, R=0. As one would expect in this case, it follows that W=0. On the other hand, if it rains on every trip the professor again stays perfectly dry since he simply carries one umbrella back and forth for the entire semester. Equation (5-52) is correct in this case as well since W=0 when R=1.

If it rains half the time, the professor will get wet on 10% of the trips. Interested readers may wish to determine the value of R that produces the largest value of W. Can this problem be solved without using differential calculus?

5.8 Interval selection and workload characterization

Observational stochastics can be regarded as the study of trajectories that have been generated by finite state automata while processing workloads that are loosely constrained. This chapter has dealt primarily with the structure of finite state automata that are used to model the behavior of actual systems. The focus has been on the identification of states, transitions and input symbols.

Workloads, which are strings of input symbols, are another integral part of observational stochastics. Workloads are characterized in terms of global parameters such as the proportion of time the walker turns to the right, the proportion of time it is raining at the start of a trip, or similar values that appear in global transition matrices. These global parameters allow the detailed structure of the workload to remain uncertain while still providing enough information to carry out an analysis.

In the majority of cases, it is both reasonable and appropriate to use global parameters that summarize aspects of an entire trajectory. Although some details are always lost during the summarization process, their loss is of no concern when these details are immaterial to the analysis. On the other hand, there are certain special cases where mathematically immaterial details are still highly significant from a practical perspective.

To examine this issue further, return to the umbrella problem discussed in the previous section. Suppose some additional information regarding weather patterns is now introduced. In particular, assume that the semester is divided into three segments. The first segment is a long dry season of 50 days when it never rains. This is followed by a short transition segment of 6 days where rainy and dry periods are interspersed. Finally, the third segment corresponds to a long stormy season of 50 days when it rains constantly.

Assume that the proportion of rainy trips during the transition segment is equal to one half. This implies that the total number of rainy trips is equal to six (from the transition segment) plus one hundred (from the stormy season). Since there are 212 trips during the entire semester, R is equal to one half.

As already noted, setting R equal to one half in equation (5-52) leads to a prediction that the professor will get wet on 10% of the trips. However, this prediction cannot possibly be correct because there are no wet trips during the dry season or the stormy season. These two segments account for $200/212 = 94.3$ % of the trips in the trajectory, which means the percentage of wet trips cannot exceed 5.7%.

The reason equation (5-52) produces an incorrect value of W is easy to understand. Equation (5-52) is based on the assumption that the observed proportion of rainy trips is empirically independent of the number of available umbrellas. In other words, $R = R_0 = R_1 = R_2$. Under the very special weather conditions just described, the assumption of empirical independence does not hold.

The actual values of R_0, R_1 and R_2 are influenced to a minor degree by the detailed structure of the transition period. For simplicity, assume that the 12 trips in the transition segment are comprised of a strictly alternating sequence of rainy and dry trips, and that the first trip in the sequence is rainy. This generates a total of six wet trips for the entire semester, which is the maximum possible number of wet trips when the transition segment is six days long.

To compute the values of R_0, R_1 and R_2 for this case, the professor's trips during each segment must be analyzed separately.

During segment 1, the professor makes 100 trips. All trips begin in state 1, and all trips are dry.

During segment 2, the professor makes 12 trips. The first rainy trip begins in state 1. This is followed by a total of six dry trips beginning in state 2 and five rainy trips beginning in state 0. All five rainy trips are wet trips.

During segment 3, the professor makes 100 rainy trips. The first trip begins in state 0 and is a wet trip. For the rest of this segment, the professor cycles between states 1 and 2, with 50 rainy trips beginning in state 2 and 49 trips beginning in state 1. The professor ends the semester in state 1, which implies that the endpoints of the trajectory are matched.

This analysis implies that a total of 6 rainy trips begin in state 0, 50 rainy trips begin in state 1, and 50 rainy trips begin in state 2. Also, the total number of trips that begin in states 0, 1 and 2 are 6, 150 and 56 respectively. Thus

$$R_0 = 6/6 = 1 \tag{5-53}$$

$$R_1 = 50/150 = 1/3 \tag{5-54}$$

$$R_2 = 50/56 = 25/28 \tag{5-55}$$

Since the proportion of rainy trips is not the same for states 0, 1 and 2, the assumption of empirical independence is violated. Thus, substituting the overall value of R (i.e., $106/212 = \frac{1}{2}$) into equation (5-52) leads to the incorrect prediction that W is equal to .10. The correct value of W in this case is $6/212 = .028$, so the error is substantial.

One of the advantages of observational stochastics is that it is always possible to carry out a more detailed analysis in such cases by simply removing the assumption of empirical independence. The balance equations associated in the global transition matrix in Figure 5-16 then become:

$$P(0) = (1 - R_2) \times P(2) \tag{5-56}$$

$$P(1) = (1 - R_1) \times P(1) + R_2 \times P(2) \tag{5-57}$$

$$P(2) = P(0) + R_1 \times P(1) \tag{5-58}$$

The values of P(j) for this more general case are then given as follows:

$$P(0) = \frac{R_1 \times (1 - R_2)}{2R_1 + R_2 - R_1 \times R_2} \tag{5-59}$$

$$P(1) = \frac{R_2}{2R_1 + R_2 - R_1 \times R_2} \tag{5-60}$$

$$P(2) = \frac{R_1}{2R_1 + R_2 - R_1 \times R_2} \tag{5-61}$$

As in the simpler case, the proportion of time the professor gets wet is once again equal to the proportion of trips that begin in state 0 when it is raining. Thus

$$W = P(0) \times R_0 \tag{5-62}$$

Since R_0 is equal to 1 in this case, the value of W can be obtained from equation (5-59) by setting R_1 and R_2 equal to the values in equations (5-54) and (5-55) respectively.

$$W = \frac{(1/3) \times (1 - 25/28)}{2 \times (1/3) + (25/28) - (1/3) \times (25/28)} = \frac{3}{106} \tag{5-63}$$

Even though this analysis is now mathematically correct, the model fails to represent the interesting and significant differences between segments 1, 2 and 3. In effect, these details are lost during the summarization process that takes place when R_0, R_1 and R_2 are evaluated.

Note that wet trips are concentrated in the middle of the semester during the transition period. The professor is likely to be extremely unhappy during this segment, even though wet trips only occur 2.8% of the time during the semester as a whole.

Analogous issues arise when modeling the performance of systems that are of considerable importance in the real world. For example, it is possible for a model of a transaction processing website to predict acceptable response time over a full 24 hour day even though response time is completely unacceptable during a few hours of every afternoon because of peak demand. In such cases the problem is not with the model itself. It is instead due to the systematic variability in demand that occurs during the 24 hour interval that has been selected to form the global transition matrix. In this case, the entire interval consists of two or more distinct segments with very different characteristics. Averaging these segments together may produce an accurate model (just as the value of W predicted by equation (5-63) is perfectly accurate). However, such results can also be highly misleading from a practical perspective.

The obvious solution in the case of the umbrella problem is to build three separate models corresponding to the three weather segments that make up the semester. All three models are based on the state transition diagram shown in Figure 5-15, but have different global transition matrices.

This approach is equally applicable to both observational stochastics and traditional stochastic modeling. It is especially natural in observational stochastics because the description of the dynamic step-by-step operation of the system being modeled (i.e., Figure 5-15) is entirely separate from the description of the workloads that generate individual trajectories.

5.9 Conventional stochastic analysis

As already noted, the umbrella problem appears as Exercise 6.6 in (Bertsekas and Tsitsiklis 2002), where it is modeled as a conventional Markov chain. The analysis provides useful insights into the difference between observational stochastics and traditional stochastic modeling. Both analyses a based on the implicit premise that the weather is unpredictable: there is no way to know in advance whether or not it will be rainy during any one of the professor's trips.

In the conventional stochastic model developed by Bertsekas and Tsitsiklis, this uncertainty is represented by assuming that the weather conditions for each successive trip (rainy or dry) can be represented by a sequence of independent, identically distributed random variables. For each of these random variables, r is the probability of a rainy trip and $1-r$ is the probability of a dry trip. In practice, the exact value of r is also unpredictable. However, the uncertainty regarding the value of r is beyond the scope of the original problem: r is simply treated as a symbolic parameter whose exact numerical value need not be specified to carry out the analysis.

This classic foundation makes it possible to model the professor's travels using a traditional Markov chain. The stepwise transition matrix for this Markov chain is shown in Figure 5-17.

$$\begin{pmatrix} 0 & 0 & 1 \\ 0 & 1-r & r \\ 1-r & r & 0 \end{pmatrix}$$

Figure 5-17. Stepwise transition matrix for the Markov model

As stated by Bertsekas and Tsitsiklis (2002), the objective of the analysis is to determine "the steady state probability that [the professor] gets wet during a commute." The steady state distribution for this Markov chain can be computed in the standard manner by determining the normalized eigenvector of the matrix in Figure 5-17. This yields:

$$\underline{P}(0) = \frac{1-r}{3-r} \qquad (5\text{-}64)$$

$$\underline{P}(1) = \frac{1}{3-r} \qquad (5\text{-}65)$$

$$\underline{P}(2) = \frac{1}{3-r} \qquad (5\text{-}66)$$

Equation (5-64) states that $\underline{P}(0)$, the steady state probability that the professor has no available umbrellas at the start of a trip, is equal to $(1-r)/(3-r)$. Also, the probability that it will be raining at the start of any trip is always equal to r. Thus, the probability that the professor gets wet during a trip is equal to $r \times (1-r)/(3-r)$.

Even though this solution has exactly the same form as the solution derived in Section 5.7, its interpretation is quite different. The observational solution in Section 5.7 applies directly to actual trajectories that are generated by the professor walking to and from school over the course of a semester. In contrast, the traditional stochastic solution is expressed in terms of the abstract notion of a steady state probability.

Bertsekas and Tsitsiklis recognize that practitioners are likely to be interested in applying their solution to actual trajectories. Thus they follow their analysis by a discussion of the "long term frequency interpretations" of their results. This discussion is similar to the material in Section 2.7. In particular, the authors state that "whenever we carry out a probabilistic experiment and generate a trajectory of the Markov chain over an infinite time horizon, the observed long term frequency with which state j is visited will ... with essential certainty... be equal to π_j." [Note: the symbol π_j is used by Bertsekas and Tsitsiklis (2002) in place of p(j). Also, the phrase "with essential certainty" actually appears in the next sentence and has been shifted forward to clarify the authors' actual intent.]

To provide a specific example of this general statement, suppose that weather conditions (rainy or dry) for the umbrella problem are actually determined at the start of each trip by the toss of a fair coin. Under this highly artificial assumption, the trajectory generated by the professor's travels can be regarded as the result of the "probabilistic experiment" cited by Bertsekas and Tsitsiklis. In this experiment, the values rainy and dry are drawn at random from a distribution where each option has a probability of ½.

Under these assumptions, there is of course no guarantee that the observed value of R will be exactly equal to ½. However, as the length of the trajectory increases, it becomes more and more likely that the observed value of R will be very close to the intrinsic value of r (i.e., ½), and that the observed value of P(j) will be very close to the steady state probability $\underline{P}(j)$. These conclusions do not hold with absolute certainty. Mathematical terms such as *almost surely*, *essentially certain* and *with probability one* are used to indicate that it is possible, but exceedingly unlikely, to encounter limiting cases where such conclusions are actually incorrect.

As already noted on several occasions, observational stochastics bypasses this difficulty entirely by dealing only with observable quantities such as R, the proportion of rainy trips, and W, the proportion of wet trips. As discussed in Chapter 3, these quantities are defined formally in terms of loosely constrained deterministic (LCD) models. The notion of an underlying intrinsic probability value such as r is not required for the analysis because stochastic processes are not involved in the formulation of the original problem. As a result, questions regarding the relationship between R and r, and between P(j) and $\underline{P}(j)$, simply never arise.

CHAPTER 6

Simple Continuous Time Models

6.1 Discrete and continuous time models

The mathematical analyses presented in Chapters 1 through 5 are based on the assumption that system behavior can be characterized by a trajectory such as the one shown in Figure 6-1. In graphs of this type, the vertical axis represents a set of observable states and the horizontal axis represents a time-ordered sequence of transition points. These transition points represent the only times at which a change of state can occur

For the discrete time processes discussed in Chapters 1 through 5, the only information available at each transition point is the new state that has just been entered. [It is possible for the new state to be the same as the current state.] Each transition point and its associated "new state" are represented in Figure 6-1 by a separate black dot.

The horizontal and vertical lines connecting the dots in Figure 6-1 help the reader visualize the shape of the trajectory, but are not essential. Similarly, the uniform spacing between each dot is also a visual aid. It has no analytic significance because the elapsed time between each pair of successive transition points is unknown and thus cannot play a role in the analysis.

If the time of each transition is also available, the discrete time trajectory in Figure 6-1 can be transformed into a continuous time trajectory of the type illustrated in Figure 6-2. In this case, the spacing between successive transition points is significant: it represents elapsed time. This additional information can be applied to the analysis of the trajectory in Figure 6-2, making it possible to determine the proportion of time the walker actually spends at each station or in each state. Expressions for the proportion of visits the walker makes to each station can also be derived, but are seldom the primary objective of the analysis.

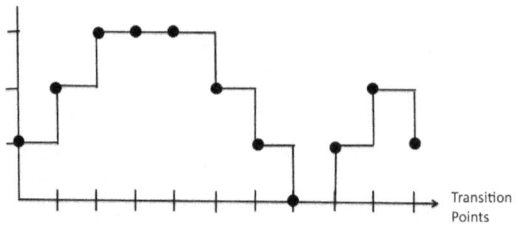

Figure 6-1. Discrete time trajectory

One other difference between discrete time and continuous time trajectories concerns the amount of time spent in the final state. In discrete time models, the last input symbol in a workload generates a transition into the final state of the trajectory. This marks the end of the trajectory as indicated in Figure 6-1. On the other hand, it is possible for a continuous time model to remain in the final state for an additional amount of time before the trajectory finally terminates. Such a case is illustrated in Figure 6-2.

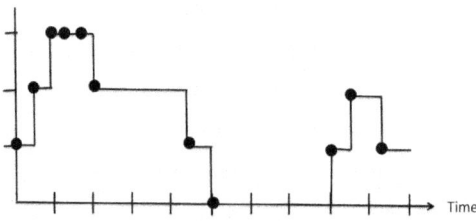

Figure 6-2. Continuous time trajectory

To accommodate this consideration, all workloads associated with continuous time LCD models must end with a special termination symbol: X. The timestamp attached to the termination symbol X represents the instant at which the trajectory actually ends, and must be greater than or equal to the timestamp of the final state in the trajectory   Tables 6-2 and 6-4 in Section 6.5 illustrate the use of the special termination symbol.

6.2 Buffer overflow in continuous time

The simple random walk discussed in Chapters 1 through 5 is closely related to a buffering process that operates in continuous time. In examples of this type, the state of the system being modeled corresponds to the number of items in a buffer rather than the position of a walker. Right and left turns correspond to increases and decreases in the number of buffered items, and rebounding off the reflecting barrier to the right of the highest state corresponds to dropping (i.e., discarding) an item that arrives when the buffer is currently full.

The next few sections describe this model in greater detail. As in the case of the random walk, the procedures used to analyze this specific example can be generalized to deal with a broad class of modeling problems. A number of these generalizations are described in Chapter 7.

6.2.1   The buffer/queue overflow problem

When requests for service arrive in a manner that is bursty rather than uniform, buffers are often provided to store incoming requests until they can be processed. Note that buffer capacity is always finite. Thus, as just noted, it may be possible for requests to arrive at a time when the buffer is already full. Such requests are typically dropped with the expectation that they will be transmitted again at some point in the future.

The designers of the ARPAnet, which is the predecessor of today's Internet, anticipated this problem. They incorporated special algorithms into the software that implements the TCP/IP protocol to trigger such retransmissions automatically if a dropped packet is suspected. [Note: For purposes of this discussion, a packet is simply a short segment of a larger message. Packets are transmitted from router to router as they make their way across a network from their origin to their ultimate destination.]

Buffer overflow is no longer a concern for the powerful routers deployed in the backbone of the modern Internet. The processing capacity of these routers is matched to the maximum bandwidth of the lines that transmit incoming packets. As a result, these routers are able to complete the processing of a packet before the final byte of the next incoming packet (on that same line) arrives. This processing takes place in parallel for each line the router supports.

Buffer overflow can still be an issue for less advanced routers. It can also bring down web servers during periods of very high load (e.g., during denial of service attacks). The next few sections examine the buffer overflow problem for the simple case of a primitive router operating in a classical store and forward network.

.6.2.2 An LCD model of buffer operation

Note that buffer operation can be represented by the state transition diagram illustrated in Figure 6-3. The number of packets in the buffer at any given instant corresponds to the state of the model. To simplify this particular example, the size of the buffer is set to an unrealistically small value of three. Extending this analysis to buffers with larger capacities is entirely straightforward.

The input symbol ⇑ corresponds to the arrival of a packet. Each time a packet arrives, the number of packets in the buffer increases by one, provided space is available. This corresponds to the state of the model increasing by one. If a packet arrives when the buffer is full (i.e., state 3), the state remains the same and the packet is dropped.

The input symbol ⇓ corresponds to the removal of a packet from the buffer after it has been transmitted to its next destination. As soon as the transmission is complete, the corresponding state in

Figure 6-3 is of course reduced by one. These step-by-step rules for operation of the buffer are entirely deterministic. However, the times between successive arrivals and the times required to transmit individual packets are usually uncertain. This uncertainty is typically represented through a stochastic model. It can also be incorporated into a continuous time LCD model of the type mentioned in Section 3.7.4.

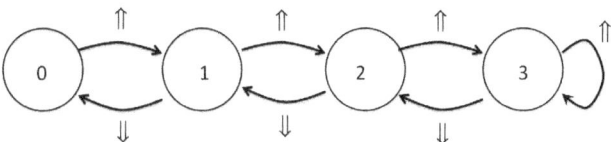

Figure 6-3. State transition diagram with input symbols

Note that discrete time and continuous time LCD models have a great deal in common. They are always based on deterministic state transition diagrams that generate trajectories by processing loosely constrained workloads. In both cases, a workload is simply an ordered sequence of input symbols. Thus, the only significant difference between discrete and continuous time models is the fact that an additional value is associated with each input symbol: the time of occurrence. Informally, this timestamp represents the instant at which the input symbol is processed and the next state in the trajectory is entered.

6.2.3 Simultaneous events

As specified in Section 3.7.4, times of occurrence are represented by sequences of real numbers that increase steadily. The input symbols that make up a workload are processed in the order in which their timestamps appear. These timestamps must form a strictly increasing sequence to rule out the possibility of two input symbols being processed at exactly the same instant.

If simultaneous events are an important consideration for a particular analysis, they can be accommodated in a continuous time LCD model by adding special input symbols to represent their appearance. For example, if two packets arrive simultaneously at a time when the number of packets in the buffer is equal to 1, the state will change immediately from 1 to 3. This can be represented by adding a transition arrow that connects state 1 to state 3. The special input symbol assigned to this arrow would then represent the simultaneous arrival of two packets.

Since there is no transition arrow connecting state 1 to state 3 in Figure 6-3, the LCD model being analyzed here implicitly excludes this particular event as well as all other events that involve simultaneous arrivals and departures. In the terminology of operational analysis, models of this type satisfy the assumption of one-step behavior (Buzen 1976b), (Denning and Buzen 1978).

6.2.4 Admissible workloads

One important difference between the state transition diagrams in Figure 6-3 and Figure 3-1 is that additional packets cannot be transmitted from the router when the buffer is empty (i.e., when the current state is state 0). As a result, the input symbol ⇓ does not appear on a transition arrow originating from state 0. This imposes a special requirement on the workloads that can be processed: workloads that cause a packet to be transmitted when in state 0 are not admissible.

The random walks analyzed in Chapters 2 and 3 have no such limitation. The walker can turn right or left after exiting from any station. Thus, for the state transition diagram in Figure 3-1, every state is assigned a transition arrow labeled with the input symbol ⇑ and a transition arrow labeled with the input symbol ⇓ .

In classical applications of finite state automata to the design of electronic circuits, all workloads must be admissible. The finite state automata employed in observational stochastics are not required to satisfy this condition.

6.2.5 Parameters

The next step is to define a set of directly observable variables that characterize the operation of a router whose buffer is represented by Figure 6-3. Following the analysis presented in (Buzen 1976b),

$A(j)^{\#}$ = number of packets that arrive at a time when the buffer contains j packets. (j = 0, 1, 2 and 3)

$C(j)^{\#}$ = number of packets whose transmission is completed at a time when the buffer contains j packets (j = 1, 2 and 3).

Note that $A(j)^{\#}$ corresponds to $R(j)^{\#}$, the number of times the walker turns to the right after leaving station j.   Similarly, $C(j)^{\#}$ corresponds to $L(j)^{\#}$, the number of times the walker turns to the left.

The next step is to define the variables A and C as follows:

A = total number of packets that arrive during the interval

$$= A(0)^{\#} + A(1)^{\#} + A(2)^{\#} + A(3)^{\#} \tag{6-1}$$

C = total number of packets that are processed during the interval

$$= C(1)^{\#} + C(2)^{\#} + C(3)^{\#} \tag{6-2}$$

The quantities $A(j)^{\#}$, $C(j)^{\#}$, A and C are total counts. In discrete time models, such totals are typically converted into proportions that can be independent of the length of the observation interval: for example, R, the proportion of turns the walker makes to the right, is equal to $[R(0)^\# + R(1)^\# + R(2)^\# + R(3)^\#] \; / \; V$.

In continuous time models, a second option is available: total counts can also be converted into rates (e.g., arrivals per second) that can once again be independent of the length of the observation interval. To convert the quantities $A(j)^\#$, $C(j)^\#$, A and C into rates, it is necessary to determine the amount of time that is spent in each state. As discussed in Section 3.7.4, these times are directly observable properties associated with continuous time trajectories of the type illustrated in Figure 6-2. To determine the total time spent in any state over an entire trajectory, first calculate the time spent in that state during each visit by subtracting the time the visit begins from the time it ends. Then sum these times over all visits made to that state during the trajectory.

Using this straightforward measurement procedure, define $T(j)^\#$ as follows for j = 0, 1, 2 and 3:

$T(j)^\#$ = the total time spent in state j

Since $T(j)^\#$ is a directly observable quantity, it is marked with a hash tag.

T = total length (i.e., duration) of the observation interval

$$= T(0)^\# + T(1)^\# + T(2)^\# + T(3)^\# \qquad\qquad (6\text{-}3)$$

B = amount of time the router is busy (actively transmitting packets from the buffer)

= amount of time the buffer contains one or more packets

$$= T(1)^\# + T(2)^\# + T(3)^\# \qquad\qquad (6\text{-}4)$$

The variables $T(j)^{\#}$, T and B can now be used to convert the total counts $A(j)^{\#}$, $C(j)^{\#}$, A and C into meaningful rates. Note first that a total of A packets arrive at the router during a total of T seconds. Thus, the overall packet arrival rate is A/T packets per second. Conventionally, the symbol λ is used to identify variables that represent arrival rates in queuing models. Accordingly, let

λ = overall arrival rate of packets at the router

$$= A / T \qquad\qquad (6\text{-}5)$$

Now consider the average amount of time required for the router to transmit a packet and remove it from the buffer. As already noted, the total number of packets transmitted during the entire interval is equal to C. Since transmissions only occur when one or more packets are in the buffer, the total amount of time the router is busy transmitting packets is equal to B. Thus the average time required to transmit an individual packet is B/C.

Let S = average amount of time required to transmit one packet

$$= B / C \qquad\qquad (6\text{-}6)$$

In the parlance of queuing theory, routers and similar devices are referred to generically as servers. The variable S represents the average service time for any such server. However, the equations of queuing theory are usually expressed in terms of service rates rather than service times. For example, if the average service time at a router is .004 seconds (i.e., 4 milliseconds), then the router is capable of transmitting a maximum of $1/.004 = 250$ packets per second. This is the average service rate of the router.

Service rates are traditionally represented by the symbol μ, which is defined as the reciprocal of S as indicated in equation (6-7).

μ = average rate at which packets are transmitted by the router when the router is active

$$= C / B$$

$$= 1 / S \tag{6-7}$$

Note that λ, S and μ reflect the observable properties of entire trajectories. It is also possible to compute these quantities during those times when the state of the system (i.e., the number of packets in the buffer) is equal to 0, 1, 2 and 3. For example, $T(2)^{\#}$ represents the total amount of time there are exactly 2 packets in the buffer, and $A(2)^{\#}$ represents the number of packets that arrive during this period. Thus $A(2)^{\#}/T(2)^{\#}$ is the rate at which packets arrive and encounter a buffer that already contains two packets. This conditional arrival rate will be denoted by $\lambda(2)$. In general:

$\lambda(j)$ = arrival rate during those times when the buffer contains j packets

$$= A(j)^{\#} / T(j)^{\#} \quad \text{for } j= 0, 1, 2 \text{ and } 3 \tag{6-8}$$

Similarly, the average rate at which packets are removed from the buffer after their transmission has been completed can also be computed for each possible state. In particular, $C(2)^{\#}/T(2)^{\#}$ is the rate at which packet transmissions are completed when the number of packets in the buffer is equal to 2. This conditional transmission rate will be denoted by $\mu(2)$. In general:

$\mu(j)$ = rate at which packets are transmitted by the router when

the buffer contains j packets

$$= C(j)^{\#} / T(j)^{\#} \quad \text{for } j= 1, 2 \text{ and } 3 \tag{6-9}$$

Since no packets can be transmitted when the buffer is empty, $C(0)^{\#}$ and $\mu(0)$ must both be zero. However, these quantities are not used in the analysis and need not be defined.

6.2.6 Homogeneity and empirical independence

In many cases of practical interest, it is reasonable to assume that the rate at which new packets arrive at a router is not affected by the number of packets already queued in that router's buffer. In other words, it is reasonable to assume that all values of $\lambda(j)$ are the same, and are thus all equal to λ as shown in equation (6-10).

$$\lambda(j) = \lambda \qquad \text{for } j = 0, 1, 2 \text{ and } 3 \tag{6-10}$$

Using the terminology introduced in Section 1.6.2, equation (6-10) states that the arrival rate of new packets is empirically independent of the state of the buffer. The term *homogeneous arrivals* was originally used in operational analysis (Buzen 1976b) to characterize the relationship expressed by equation (6-10). The term empirical independence extends this useful concept to a more general class of relationships and is thus preferred for use in observational stochastics.

Similar comments apply to service completion rates. If all values of $\mu(j)$ are the same, they are empirically independent of j and must all be equal to μ as shown in equation (6-11)

$$\mu(j) = \mu \qquad \text{for } j = 1, 2 \text{ and } 3 \tag{6-11}$$

The relationship expressed by equation (6-11) was originally referred to as the assumption of homogenous service rates. Once again, the term empirical independence generalizes the concept of homogeneous arrivals and is thus preferred for use in observational stochastics.

6.2.7 Solution using global and local balance

The next step in analyzing the buffer overflow problem is to determine the proportion of time that the number of packets in the buffer is equal to 0, 1, 2 or 3 (i.e., the proportion of time spent in states 0, 1, 2 and 3). Equation (6-12) defines these quantities.

The use of a lower case p in equation (6-12) identifies p(j) as a time-based proportion. This convention makes it possible to distinguish P(j), the proportion of visits to state j, from p(j), the proportion of time spent in state j.

p(j) = proportion of time there are j packets in the buffer

   = proportion of time spent in state j

   $= T(j)^{\#} / T$ j = 0, 1, 2 and 3   (6-12)

The analysis of the buffer overflow model depicted in Figure 6-3 follows the same approach employed in Chapters 3, 4 and 5. Lemma 4.1 is once again applicable: the number of entrances into each state must be equal to the number of exits from that state, provided the endpoints of the trajectory being analyzed are matched.

[As in the discrete time case, the assumption of matched endpoints is not essential. It is introduced to reduce the complexity of the analysis. Section 7.4 describes a general procedure for analyzing continuous time trajectories with unmatched endpoints.]

The application of Lemma 4.1 to the buffer overflow problem is easy to visualize with the help of Figure 6-4, which has the same structure as Figure 6-3. However, the label attached to each transition arrow now represents the number of times the associated transition occurs over the course of a trajectory. For example, $A(0)^{\#}$ is the total number of transitions out of state 0 and $C(1)^{\#}$ is

the total number of transitions in.   Equating transitions in and
out of each state leads directly to equations (6-13) through (6-16).

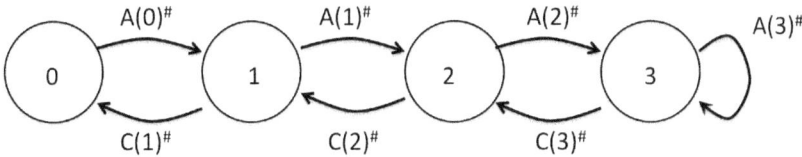

Figure 6-4. State transition diagram with transition counts

$$A(0)^{\#} = C(1)^{\#} \tag{6-13}$$

$$A(1)^{\#} + C(1)^{\#} = A(0)^{\#} + C(2)^{\#} \tag{6-14}$$

$$A(2)^{\#} + C(2)^{\#} = A(1)^{\#} + C(3)^{\#} \tag{6-15}$$

$$A(3)^{\#} + C(3)^{\#} = A(2)^{\#} + A(3)^{\#} \tag{6-16}$$

The next step is to divide both the left and the right sides of
equations (6-13) – (6-16) by T.   Then apply equations (6-17) and
(6-18).   These steps transform equations (6-13) – (6-16) into
equations (6-19) – (6-22).

$$\frac{A(j)^{\#}}{T} = \frac{A(j)^{\#}}{T(j)^{\#}} \times \frac{T(j)^{\#}}{T}$$

$$= \lambda(j) \times p(j) \quad \text{for } j = 0, 1, 2 \text{ and } 3 \tag{6-17}$$

$$\frac{C(j)^{\#}}{T} = \frac{C(j)^{\#}}{T(j)^{\#}} \times \frac{T(j)^{\#}}{T}$$

$$= \mu(j) \times p(j) \quad \text{for } j=1, 2 \text{ and } 3 \tag{6-18}$$

$$\lambda(0) \times p(0) = \mu(1) \times p(1) \tag{6-19}$$

$$\lambda(1) \times p(1) + \mu(1) \times p(1) = \lambda(0) \times p(0) + \mu(2) \times p(2) \tag{6-20}$$

$$\lambda(2) \times p(2) + \mu(2) \times p(2) = \lambda(1) \times p(1) + \mu(3) \times p(3) \tag{6-21}$$

$$\lambda(3) \times p(3) + \mu(3) \times p(3) = \lambda(2) \times p(2) + \lambda(3) \times p(3) \tag{6-22}$$

Equations (6-19) through (6-22) represent the conventional balance equations for this model. Figure 6-5 provides a convenient way to visualize these equations. This figure is obtained from Figure 6-4 by replacing raw counts (i.e., $A(j)^{\#}$ and $C(j)^{\#}$) by corresponding transition rates (i.e., $\lambda(j)$ and $\mu(j)$). Note that the definitions of $\lambda(j)$ and $\mu(j)$ are provided by equations (6-8) and (6-9).

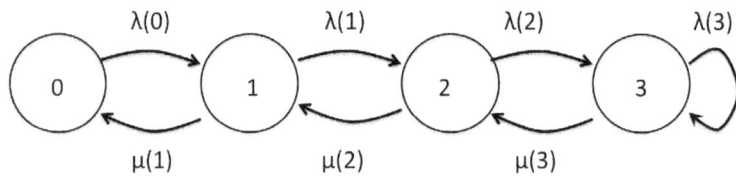

Figure 6-5. State transition diagram with transition rates

In the parlance of queuing theory, equations (6-19) through (6-22) are referred to as balance equations. or, more precisely, as *global balance* equations They represent the fact that the total rate of transition flow into each state is equal to the total number of transition flow out.

Figure 6-4 also supports a second set of balance equations that are simpler, but yield precisely the same formal solution in this case. These alternate balance equations are based on the concept of *local balance.*

Local balance is an easy concept to understand. Consider any state transition diagram of the type illustrated in Figure 6-4. Imagine that the diagram is divided into two disjoint parts: Part A and Part B. If the endpoints of a trajectory generated by this state transition diagram are matched, it is clear that the number of transitions from Part A to Part B must be equal to the number of transitions from Part B to Part A. Applying this logic to a partitioning boundary that separates state 0 from state 1 produces equation (6-23). Similarly, a partitioning boundary that separates state 1 from state 2 produces equation (6-24), and a partitioning boundary that separates state 2 from state 3 produces equation (6-25).

$$A(0)^{\#} = C(1)^{\#} \tag{6-23}$$

$$A(1)^{\#} = C(2)^{\#} \tag{6-24}$$

$$A(2)^{\#} = C(3)^{\#} \tag{6-25}$$

The same steps that transformed equations (6-13) through (6-16) into equations (6-19) through (6-22) can now be applied to equations (6-23) through (6-25). The resulting local balance equations are given by equations (6-26) through (6-28).

$$\lambda(0) \times p(0) = \mu(1) \times p(1) \tag{6-26}$$

$$\lambda(1) \times p(1) = \mu(2) \times p(2) \tag{6-27}$$

$$\lambda(2) \times p(2) = \mu(3) \times p(3) \tag{6-28}$$

Note that these local balance equations are substantially simpler than the corresponding global balance equations. In addition, the linkage between equations (6-26) through (6-28) and the state transition diagram in Figure 6-7 is direct and easy to understand.

In this particular case, equation (6-26) leads immediately to equation (6-30). Similarly, equation (6-27) can be combined with equation (6-30) to generate equation (6-31). Also, equation (6-28) can be combined with equation (6-31) to generate equation (6-32). Combining these three equations with the standard normalization constraint then yields equation (6-29).

The global balance equations specified by equations (6-19) through (6-22) yield exactly the same solution. The standard normalization constraint is again required because only three of the four global balance equations are linearly independent.

$$p(0) = \frac{1}{1 + \dfrac{\lambda(0)}{\mu(1)} + \dfrac{\lambda(0) \times \lambda(1)}{\mu(1) \times \mu(2)} + \dfrac{\lambda(0) \times \lambda(1) \times \lambda(2)}{\mu(1) \times \mu(2) \times \mu(3)}} \qquad (6\text{-}29)$$

$$p(1) = \frac{\lambda(0)}{\mu(1)} \times p(0) \qquad (6\text{-}30)$$

$$p(2) = \frac{\lambda(0) \times \lambda(1)}{\mu(1) \times \mu(2)} \times p(0) \qquad (6\text{-}31)$$

$$p(3) = \frac{\lambda(0) \times \lambda(1) \times \lambda(2)}{\mu(1) \times \mu(2) \times \mu(3)} \times p(0) \qquad (6\text{-}32)$$

For the case where the values of $\lambda(j)$ are equal to λ for $j = 0, 1, 2$ and 3, and where $\mu(j) = \mu$ for $j = 1, 2$ and 3, the local balance equations reduce to equations (6-33) through (6-35).

$$\lambda \times p(0) = \mu \times p(1) \qquad (6\text{-}33)$$

$$\lambda \times p(1) = \mu \times p(2) \qquad (6\text{-}34)$$

$$\lambda \times p(2) = \mu \times p(3) \tag{6-35}$$

The normalized solution to these equations is:

$$p(0) = \frac{1}{\displaystyle\sum_{j=0}^{3} (\lambda/\mu)^j} \tag{6-36}$$

$$p(1) = \frac{\lambda/\mu}{\displaystyle\sum_{j=0}^{3} (\lambda/\mu)^j} \tag{6-37}$$

$$p(2) = \frac{(\lambda/\mu)^2}{\displaystyle\sum_{j=0}^{3} (\lambda/\mu)^j} \tag{6-38}$$

$$p(3) = \frac{(\lambda/\mu)^3}{\displaystyle\sum_{j=0}^{3} (\lambda/\mu)^j} \tag{6-39}$$

Generalized variants of equations (6-29) through (6-32) and equations (6-36) through (6-39) were presented as part of the original foundation for operational analysis (Buzen 1976b).

6.2.8 Proportion of dropped packets

Recall that the primary goal of this analysis is to determine the proportion of packets that are dropped: that is, the proportion of packets that arrive at a time when the buffer already contains three packets and is thus completely full. This proportion must be expressed as a function of $\lambda(j)$ and $\mu(j)$. To begin this phase of the analysis, first define $\vec{p}(j)$ as follows:

$\vec{p}(j)=$ proportion of packets that find j packets already in the buffer when they arrive

$$= \frac{A(j)^{\#}}{A(0)^{\#}+A(1)^{\#}+A(2)^{\#}+A(3)^{\#}} = \frac{A(j)^{\#}}{A} \qquad (6\text{-}40)$$

Equation (6-40) can be rewritten as follows:

$$\vec{p}(j) = \frac{A(j)^{\#}}{T(j)^{\#}} \times \frac{T(j)^{\#}}{T} \times \frac{T}{A} \qquad (6\text{-}41)$$

Applying equations (6-5), (6-8), and (6-12):

$$\vec{p}(j) = \frac{\lambda(j)}{\lambda} \times p(j) \qquad (6\text{-}42)$$

Equation (6-42) expresses a fundamental relationship between the queue length seen by an arriving packet and the queue length seen by an external observer who computes the overall proportion of time that the queue is in each possible state. This relationship is an *operational law* that is valid for all trajectories, including those whose endpoints are not matched. Only $\vec{p}(3)$ is of interest here since it represents the proportion of packets that are dropped because they arrive at a time when the buffer is already full.

To complete the solution of the buffer overflow problem, the value of $p(3)$ that appears on the right hand side of equation (6-42) for $j=3$ must be expressed as a function of the parameters $\lambda(j)$ and $\mu(j)$. This can be done by combining equations (6-29) and (6-32) with equation (6-42) to obtain equation (6-43):

$$\vec{p}(3) = \frac{\lambda(3)}{\lambda} \times \frac{\lambda(0)\lambda(1)\lambda(2)}{\mu(1)\mu(2)\mu(3)} \times \frac{1}{1 + \dfrac{\lambda(0)}{\mu(1)} + \dfrac{\lambda(0)\lambda(1)}{\mu(1)\mu(2)} + \dfrac{\lambda(0)\lambda(1)\lambda(2)}{\mu(1)\mu(2)\mu(3)}}$$

$$(6\text{-}43)$$

Equation (6-43) expresses the proportion of dropped packets for a buffer of size three. It is valid for any trajectory with matched endpoints. No other assumptions are required.

Equation (6-43) can be simplified substantially if it is assumed that all values of $\lambda(j)$ that appear in this equation are equal to λ and all values of $\mu(j)$ that appear in this equation are equal to μ. Making these same simplifying assumptions in the general case where the size of the buffer is equal to N yields the following solution for $\vec{p}(N)$, the proportion of dropped packets when buffer size is equal to N:

$$\vec{p}(N) = \frac{(\lambda/\mu)^N}{\displaystyle\sum_{j=0}^{N} (\lambda/\mu)^j} \qquad (6\text{-}44)$$

To illustrate how equation (6-44) can be applied in practice, suppose the designer of a router must decide upon the size of the internal buffer. The router is capable of transmitting packets at a maximum rate of 250 packets per second. This hardware-based parameter is represented in equation (6-44) by μ.

Suppose the router is being designed to handle a load of 200 packets per second with a success rate of 99.9% (i.e., with the proportion of dropped packets less than .001). Equation (6-44) can be used to determine the minimum buffer size that must be specified to meet this objective, assuming that packet arrival rates and packet transmission rates are both empirically independent of the number of packets in the buffer.

To apply equation (6-44), simply set μ equal to 250 and λ equal to 200. Then increase N until $\bar{p}(N)$ drops below .001. Table 6-1 presents the results of these computations. The first row presents a series of buffer sizes, and the second presents the corresponding proportion of dropped packets.

N	3	6	9	12	15	18	21	24
$\bar{p}(N)$	.173	.066	.030	.015	.007	.004	.002	.001

Table 6-1. Proportion of dropped packets as a function of buffer size

As shown in Table 6-1, setting the buffer size to 24 reduces the proportion of dropped packets to less than 0.001 and meets the design goal of a 99.9% success rate. The computations shown in Table 6-1 will be exactly correct for any trajectory that satisfies the assumptions of the analysis: the endpoints must be matched, and the values of $\lambda(j)$ and $\mu(j)$ must be empirically independent of the number of packets in the buffer. The detailed structure of these trajectories is immaterial to the analysis and can remain uncertain. This is of course typical of the results derived using observational stochastics.

As discussed in Section 7.4, the assumption of matched endpoints usually has a negligible impact on such computations when the trajectory is at least moderately long. In addition, the assumption that the values of $\lambda(j)$ are empirically independent of the number of packets in the buffer is usually appropriate. If the values of $\lambda(j)$ are all close to, but not exactly equal to 200, a sensitivity analysis of the type outlined in Section 1.6.4 can always be carried out. This analysis requires generalized extensions of equations (6-36) and (6-39) that are given by equations (6-45) and (6-46).

$$p(0) = \frac{1}{1 + \displaystyle\sum_{k=1}^{N} \prod_{j=1}^{k} \frac{\lambda(j-1)}{\mu(j)}} \tag{6-45}$$

$$p(N) = p(0) \times \prod_{j=1}^{N} \frac{\lambda(j-1)}{\mu(j)} \tag{6-46}$$

The assumption that the values of $\mu(j)$ are empirically independent of the number of packets in the buffer is appropriate in some cases but not others. Even when this assumption is not entirely appropriate, it often yields a conservative estimate of the minimum buffer size required to obtain satisfactory performance. This property makes equation (6-44), and other equations based on similar assumptions, very useful in practice. Once again, equations (6-45) and (6-46) can be used to carry out a sensitivity analysis in cases where the assumption of empirical independence is approximately (but not exactly) correct.

6.3 Continuous time Markov models

All problems discussed in this chapter have been formulated and analyzed within the context of observational stochastics. These problems can also be treated using traditional stochastic models. This section, which is intended primarily for students, provides an informal introduction to these models. Advanced readers who are already familiar with stochastic modeling may wish to skip this material and proceed to Section 6.4.

6.3.1 Traditional representation of variability

The buffer overflow problem introduced in Section 6.2 is traditionally analyzed using a continuous time Markov model. At its core, this model is based on the same state transition diagram shown in Figure 6-3. However, the mechanisms that determine the time and the direction of each transition are entirely different in this case, even though the ultimate conclusions of the analysis are the same.

The motivation for the stochastic formulation is easy to appreciate. Consider the numerical example used in Section 6.2.9. Packets arrive at the buffer at a rate of 200 per second, implying that the average time between arrivals is 5 milliseconds. The rate at which packets can be transmitted by the router and removed from the buffer is 250 per second, implying that the average transmission time per packet is 4 milliseconds. If each inter-arrival time is exactly 5 milliseconds in length and each transmission time requires exactly 4 milliseconds, and if the observation interval begins with an empty buffer, each arriving packet will be transmitted and removed from the buffer before the next packet arrives. The number of packets in the buffer will simply alternate between zero and one, and packets will never be dropped.

While this analysis is entirely valid, it is based on assumptions that are not likely to be encountered in practice. In particular, the assumption that packets arrive at evenly spaced intervals exactly 5 milliseconds apart is unrealistic for routers functioning in real networks where arriving packets are generated by a large set of end users who operate in an independent manner. Under such conditions, variability is always present and there is always a chance that buffers will overflow and packets will be dropped.

One way to represent this type of variability is to regard each successive inter-arrival time as a sample from a pre-specified probability distribution. This assumption forms the basis for a traditional stochastic model that can be solved analytically or evaluated numerically using Monte Carlo simulation. For purposes of this discussion, assume that a simulation program is being employed. It is clear that the probability distribution used to represent the inter-arrival times will influence the observed values of $\lambda(j)$ that are extracted from the output of the simulation.

6.3.2 Motivation for the exponential distribution

Note that $\lambda(j)$ represents the rate at which new packets arrive when the number of packets already in the buffer is equal to j. If the number of packets that are being transmitted through the network is large, it is reasonable to assume that the arrival rate at the router will be independent of the number of packets already waiting in the buffer. In other words, the arrival rate will neither speed up nor slow down as the number of buffered packets varies, making it reasonable to assume that the values of $\lambda(j)$ will all be the same.

This raises an interesting mathematical question: what distribution of inter-arrival times yields values of $\lambda(j)$ that, in the long run, will all be the same? The only probability distribution that satisfies this objective under very general conditions is the exponential distribution (also referred to as the negative exponential distribution).

The form of this distribution is specified by equation (6-47), where α is the sole parameter and e is a standard mathematical constant: approximately 2.71828). The mean of this distribution is $1/\alpha$.

$$\text{Probability [inter-arrival time} \leq t] = 1 - e^{-\alpha t} \qquad (6\text{-}47)$$

Exponential distributions have profoundly important mathematical properties. In the context of this discussion, only one property is significant: whenever the distribution specified by equation (6-47) is used to generate inter-arrival times in a Monte Carlo simulation, the values of $\lambda(j)$ extracted from the output of the simulation will almost surely be the same in the long run, and all be equal to $1/\alpha$. This ability of exponential distributions to generate observable behavior that conforms to the intuitive expectations of practitioners is the key to their practical value.

One minor comment on terminology: when inter-arrival times are characterized by an exponential distribution of the form specified in equation (6-47), arrivals are said to be generated by a Poisson process. The term Poisson arrivals is also used in such cases.

6.3.3 Traditional derivation of the steady state distribution

To complete the specification of the traditional stochastic model of the buffer overflow problem, it is also necessary to characterize the amount of time required to transmit each packet. Once again, it is conventional to assume that each individual transmission time can be regarded as an independent sample from another exponential distribution.

These probabilistic assumptions regarding inter-arrival times and service times imply that the number of packets in the buffer at any time t can be represented by a random variable whose associated probability distribution is defined as follows:

$\underline{p}(j,t)$ = probability that there are j packets in the buffer at time t.

It is clear that $\underline{p}(j,t)$ depends on the number of packets in the buffer when $t = 0$ and the probability distributions that characterize inter-arrival times and packet transmission times. As t grows large, the dependence of $\underline{p}(j,t)$ on the initial state diminishes and, in the limit, disappears entirely. More specifically, each value of

$\underline{p}(j,t)$ approaches a steady state or stationary probability $\underline{p}(j)$ as t approaches infinity.   This is directly analogous to the limiting behavior of the discrete time random walk discussed in Section 2.4 and illustrated in Table 2-1. The main difference is that time now advances continuously rather than in discrete steps. Thus, time must be represented by a real number rather than an integer.

Although the extension from discrete time to continuous time may seem relatively minor, the complexity of the mathematical analysis increases substantially when time flows continuously. Recall that the steady state distribution of a discrete time Markov process is obtained by setting the distribution after step $t+1$ equal to the distribution after step t. This produces a set of balance equations that can be solved to determine the steady state distribution.

This argument cannot be applied directly to a continuous time Markov process because the probability distribution $\underline{p}(j,t)$ evolves smoothly in steps that are infinitesimally small. Instead of dealing with distributions at time t and time $t+1$, it is necessary to deal with distributions at time t and time $t+\varepsilon$ where ε is arbitrarily small.

On an informal intuitive level, it is sometimes useful to regard these arbitrarily small steps as being driven by flows that move probability mass smoothly from one state to another. When a continuous time Markov process is in steady state, the overall rate at which probability mass flows into each state is equal to the rate at which it flows out. This implies that the rate at which the probability distribution is changing is equal to zero.   This intuitively appealing condition can be expressed in a mathematically rigorous form by setting the first derivative of the probability distribution equal to zero. This generates a set of balance equations whose solution is the steady state distribution of the underlying stochastic process.

For most practitioners, the formidable mathematical arguments that underlie this analysis can be safely ignored. The only relevant point is that the balance equations that determine the steady state distribution of a continuous time Markov process have exactly the same form as the balance equations that determine the corresponding values of $p(j)$ when observational stochastics is employed. Assuming that inter-arrival times and packet transmission times are exponentially distributed, the balance equations for the traditional Markov model of the buffer overflow problem are given by equations (6-48) through (6-51).

$1/\alpha$ = mean of the exponentially distributed random variable that characterizes the time between successive arrivals at the buffer

$1/\beta$ = mean of the exponentially distributed random variable that characterizes the time to transmit a packet

$\underline{p}(j)$ = steady state (stationary) probability that there are j packets in the buffer

$$\alpha \times \underline{p}(0) = \beta \times \underline{p}(1) \tag{6-48}$$

$$\alpha \times \underline{p}(1) + \beta \times \underline{p}(1) = \alpha \times \underline{p}(0) + \beta \times \underline{p}(2) \tag{6-49}$$

$$\alpha \times \underline{p}(2) + \beta \times \underline{p}(2) = \alpha \times \underline{p}(1) + \beta \times \underline{p}(3) \tag{6-50}$$

$$\beta \times \underline{p}(3) = \alpha \times \underline{p}(2) \tag{6-51}$$

These balance equations have exactly the same form as equations (6-19) through (6-22). However, the derivation of equations (6-19) through (6-22) is based on the simple idea that the number of entrances into each state is equal to the number of exits out. Differential equations play no role in the formal characterization of the mathematical model, and there is no need to set derivatives equal to zero to obtain the balance equations that lead ultimately to the desired solution. This is yet another benefit of observational stochastics: the complexity of the analysis does not increase when dealing with systems that operate in continuous, rather than discrete, time.

Because the balance equations are the same, the observational model and the traditional stochastic model lead to exactly the same practical conclusions. The details of the derivation need not be considered here. All that matters here is that the classical stochastic solution to the buffer overflow problem with Poisson arrivals and exponential service time is given by equations (6-52) through (6-56).

$$\underline{p}(0) = \frac{1}{\displaystyle\sum_{j=0}^{3} (\alpha/\beta)^j} \tag{6-52}$$

$$\underline{p}(1) = \frac{\alpha/\beta}{\displaystyle\sum_{j=0}^{3} (\alpha/\beta)^j} \tag{6-53}$$

$$\underline{p}(2) = \frac{(\alpha/\beta)^2}{\displaystyle\sum_{j=0}^{3} (\alpha/\beta)^j} \tag{6-54}$$

$$\underline{p}(3) = \frac{(\alpha/\beta)^3}{\displaystyle\sum_{j=0}^{3} (\alpha/\beta)^j} \tag{6-55}$$

$\bar{p}(3)$ = probability a packet is dropped because it arrives when the buffer is full

$$= \underline{p}(3)$$

$$= \frac{(\alpha/\beta)^3}{\displaystyle\sum_{j=0}^{3} (\alpha/\beta)^j} \tag{6-56}$$

6.4 Comparing traditional and observation models

Even though the classical stochastic results in equations (6-52) through (6-56) and the corresponding observational results in equations (6-36) through (6-39) have the same algebraic form, the fundamental differences discussed in Chapters 1, 2 and 3 remain. The structural considerations that lead to these differences are illustrated in Figure 6-6.

As indicated by the middle box in the top row of Figure 6-6, both modeling approaches are based upon the assumption that the step-by-step operation of the system being modeled can be represented by a deterministic state transition diagram of the type illustrated in Figure 6-3. However the two approaches employ different methods for characterizing the uncertainty and variability associated with each individual transition that takes place over the course of a trajectory.

Observational stochastics is based on the assumption that the workloads and trajectories associated with a state transition diagram satisfy a set of loose constraints: in this example, all endpoints must be matched and the values of $\lambda(j)$ and $\mu(j)$ must be empirically independent of j. The state transition diagram in Figure 6-3, together with these loose constraints, forms a loosely constrained deterministic (LCD) model that is sufficient for the derivation of equations (6-36) through (6-39).

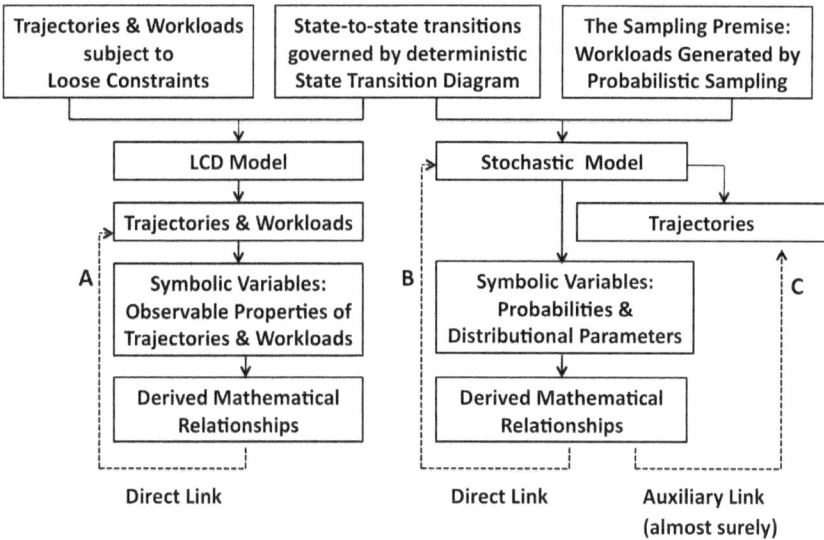

Figure 6-6. LCD models versus stochastic models

Note that all symbolic variables used in the specification and analysis of an LCD model are defined in terms of directly observable properties of the workloads and trajectories associated with that model (e.g., λ, μ and p(j) are defined by equations (6-5), (6-7) and (6-12)). This implies that all mathematical relationships derived from the model apply directly to these trajectories and workloads. Dotted line A in Figure 6-6 reflects this direct linkage.

The construction of a classical stochastic model follows a different path that is depicted on the right side of Figure 6-6. For the buffer overflow model in Section 6.3, it is assumed that each inter-arrival time and each packet transmission time can be regarded as a sample from an underlying probability distribution. This assumption, which has been referred to previously as the sampling premise, leads to the formulation of a stochastic model: that is, a traditional stochastic process that is being used as a model of a real or hypothetical system.

Note that the symbolic variables associated with a stochastic model are not defined in terms of observable properties of the model's workloads and trajectories. As in the discrete time case, these symbolic variables characterize probability distributions that are associated with the underlying stochastic process. Thus all mathematical relationships derived from the analysis of a stochastic model apply directly to the underlying stochastic process. This linkage is represented by dotted line B.

For the traditional stochastic model of buffer overflow discussed in Section 6.3, the symbolic variables α and β are parameters of exponential distributions. These distributions are incorporated into the formal specification of the underlying stochastic process. Also, the symbolic variables $\underline{p}(j)$ for $j = 0, 1, 2$ and 3 represent the steady state distribution of the process itself. Note that these symbolic variables represent quantities that are quite different from the directly observable quantities represented by λ, μ and $p(j)$.

An analysis that begins with a rigorous specification of a stochastic process and proceeds to derive mathematical relationships among the parameters and probability distributions of that process is, in essence, an exercise in pure mathematics. However, the mathematical relationships derived in such cases also have important practical applications: they can be linked by the Ergodic Theorem to the observable properties of trajectories associated with the underlying stochastic process. This auxiliary linkage, which is depicted by dotted line C, is separate from the primary flow of the mathematical analysis.

6.5 Application of the observational solution

Setting N equal to three in equation (6-44) yields the proportion of dropped packets for buffers of capacity three. This solution is exactly correct for any trajectory with matched endpoints, provided the values of $\lambda(j)$ and $\mu(j)$ are empirically independent of j. To illustrate an application of this result, two trajectories with very different shapes will now be considered.

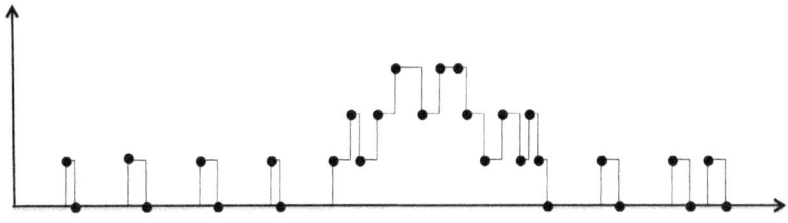

Figure 6-7. Trajectory satisfying the assumptions of equation (6-44)

⇑	⇓	⇑	⇓	⇑	⇓	⇑	⇓	⇑	⇑	⇓	⇑	⇑	⇓	⇑
8	10	18	21	29	32	40	42	50	53	55	58	61	65	68
⇑	⇓	⇓	⇑	⇓	⇑	⇓	⇓	⇑	⇓	⇑	⇓	⇑	⇓	X
71	72	75	78	81	83	85	87	95	98	106	109	112	115	120

Table 6-2. Workload for trajectory in Figure 6-7

The trajectory shown in Figure 6-7 is generated when the state transition diagram in Figure 6-3 processes the workload shown in Table 6-2. Note that the arrival and departure times that appear in rows two and four of Table 6-2 are in milliseconds. The entire interval is 120 milliseconds in length. Each input symbol is marked by a dot placed along the trajectory in Figure 6-7 at the appropriate point.

# in buffer	$T(j)^{\#}$	$A(j)^{\#}$	$C(j)^{\#}$	$\lambda(j) =$ $A(j)^{\#}/T(j)^{\#}$	$\mu(j)$ $=C(j)^{\#}/T(j)^{\#}$	$P(j) =$ $T(j)^{\#}/T$
j=0	64 ms	8	-	125/sec	--	64/120
j=1	32 ms	4	8	125/sec	250/sec	32/120
j=2	16 ms	2	4	125/sec	250/sec	16/120
j=3	8 ms	1	2	125/sec	250/sec	8/120
Total	120	15	14			

Table 6-3. $\lambda(j)$ and $\mu(j)$ for trajectories in Figures 6-7 and 6-8

Since the buffer is empty at the beginning and end of the interval, the endpoints of the trajectory are matched. Also, as shown in Table 6-3, the values $\lambda(j)$ are all equal to 125 packets per second and the values of $\mu(j)$ are all equal to 250 packets per second. Thus, the assumptions required to justify the application of equation (6-44) are satisfied, and the link depicted by dotted line A in Figure 6-7 is applicable.

Figure 6-8 presents a different trajectory that satisfies this same set of assumptions. This trajectory is generated when the state transition diagram in Figure 6-3 processes the workload in Table 6-4. The trajectory in Figure 6-8 is also 120 milliseconds in length, and its endpoints are once again matched. Even though the trajectories have different shapes, the values in Table 6-3 are exactly the same in both cases. Thus, the loose constraints required to ensure the correctness of equation (6-44) are again satisfied, and the link depicted by dotted line A in Figure 6-6 is valid for this trajectory as well.

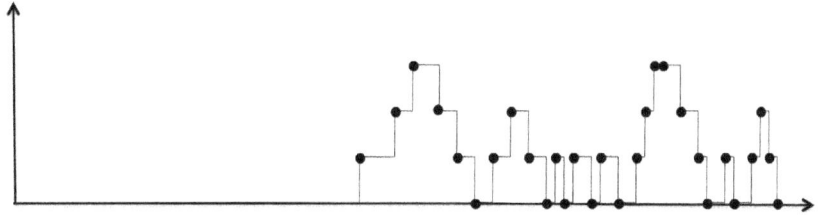

Figure 6-8. Another trajectory satisfying the assumptions of equation (6-44)

⇑	⇑	⇑	⇓	⇓	⇓	⇑	⇑	⇓	⇓	⇑	⇓	⇑	⇓	⇑
46	51	54	58	61	64	67	70	73	76	78	80	82	85	87
⇓	⇑	⇑	⇑	⇑	⇓	⇓	⇓	⇑	⇓	⇑	⇑	⇓	⇓	X
90	93	95	97	98	101	104	106	109	111	114	116	118	120	120

Table 6-4. Workload for trajectory in Figure 6-8

For cases where $\lambda = 125$ and $\mu = 250$, equation (6-44) implies that the proportion of dropped packets is equal to $1/15 = 6.7\%$. To verify the correctness of this prediction, note that 15 packets arrive during the course of each trajectory. In each case, exactly one packet is dropped. The dropped packet in Figure 6-7 arrives 71 milliseconds after the start of the interval, and the dropped packet in Figure 6-8 arrives 98 milliseconds after the start of the interval. Thus, equation (6-44) yields the correct solution for both trajectories.

Note that the analysis does not attempt to predict the exact time at which a packet will be dropped. Instead, only the proportion of dropped packets is predicted. The actual number of packets in the buffer at any specific point along the trajectory is immaterial to the analysis and can remain uncertain.   As mentioned earlier, this form of uncertainty does not depend upon the sampling premise. There is no need to assume that the values in Table 6-2 and Table 6-4 have been drawn from probability distributions.

6.6 Application of the stochastic solution

Dotted line C in Figure 6-6 represents the connection between a stochastic model and a set of associated trajectories. To understand the nature of this connection, recall that a stochastic process can be regarded as an ordered sequence of probability distributions that all share the same sample space. Intuitively, the probability distribution associated with a particular instant of time specifies the probability of the stochastic process being in each possible state at that instant. If the stochastic process is in steady state, these probabilities remain stationary (i.e., fixed) as time moves forward.

As noted at the end of Section 6.3, equations (6-52) through (6-55) represent the steady state distribution of the traditional stochastic model of buffer overflow. The symbolic variables α and β that appear in these equations are parameters of exponential distributions that characterize inter-arrival times and packet transmission times in the underlying stochastic process. Dotted line B that appears in Figure 6-6 represents the linkage between equations (6-52) through (6-55) and this mathematical model.

On an informal level, the trajectories that are linked to a stochastic model in Figure 6-6 can be thought of as the output of all possible Monte Carlo simulations of the underlying stochastic process. For the buffer overflow problem, imagine that the simulation implements the step-by-step logic in Figure 6-3 and that inter-arrival times and packet transmission times are determined by sampling from exponential distributions with parameters α and β respectively. Each successive run of the simulation generates a different trajectory.

The Ergodic Theorem applies to the limiting properties of all possible trajectories that such a simulation can generate. The theorem states that the observed proportion of time that the buffer actually contains j packets will, in the limit, almost surely converge to the steady state probability $\underline{p}(j)$. These steady state probabilities are given by equations (6-52) through (6-55) in this particular case.

This linkage between the steady state probability distribution of a stochastic process and the observable properties of its associated trajectories is indicated by dotted line C in Figure 6-6. The formal derivation of equations (6-52) through (6-55) does not depend in any way on the existence of this linkage. However, it is critical to the practical application of these results.

Note that the Ergodic Theorem makes no assertions regarding the properties of trajectories whose length is finite. In contrast, the solutions derived through observational stochastics (e.g., equation (6-44)) apply with complete certainty to all finite length trajectories that satisfy the assumptions of the LCD model being analyzed.

6.7 Verification, validation and risk

Practitioners who apply models to problems that arise in the real world have an obvious interest in verifying that the assumptions incorporated into their models are correct and that the predictions generated by their models are accurate. However, even after a model has been carefully validated, there is always a risk it will generate an incorrect prediction when applied to a new situation.

These risks can be grouped into three broad categories: modeling risk, forecasting risk, and sampling risk. The next few sections discuss each category separately and examine a number of issues that are applicable to a broad class of mathematical and simulation models.

6.7.1 Modeling risk for LCD models

All models incorporate assumptions regarding the nature of the real or hypothetical systems they are intended to represent. Modeling risk is simply the risk that the system being modeled may not satisfy the assumptions that underlie the model being employed. The issue of modeling risk is especially challenging in cases where the system being modeled does not yet exist since modeling assumptions cannot be tested directly. In such cases, practitioners must rely on their judgment that the model provides an adequate representation of the factors and interactions that influence the behavior of the system being modeled.

One way to build confidence in any model is to conduct a series of thought experiments to determine whether the model's predictions agree with one's intuitive expectations of how the system is likely to perform. For example, in the case of the buffer overflow model, it is reasonable to expect the percentage of dropped packets to decrease if the size of the buffer is increased or if the rate at which packets arrive is reduced.

A more interesting thought experiment is to examine the effect of reducing both the arrival rate and the size of the buffer by a factor of two. This is equivalent to replacing a single router that has a buffer of size N by two routers, each with a buffer size of N/2 and each processing exactly half the load of the original router. Readers are encouraged to think carefully about the likely impact of this change before seeing if their intuition agrees with the solution that is implicit in equation (6-44).

Once a model has passed the practitioner's tests of reasonableness, the next step is to verify the assumptions of the model and validate its predictions. This is, of course, only possible if the system being modeled actually exists. As illustrated in Figure 6-9, an LCD model is comprised of two main components: a state transition diagram and a set of loose constraints on the workloads and trajectories associated with that diagram. In this case, modeling risk arises because the system being modeled may not conform to the assumptions inherent in either or both of these components.

Experiments that are capable of assessing this risk are easy to imagine. Simply observe the behavior of the system over one or more intervals of time and measure both the external workload driving the system and the trajectory that is actually followed when this workload is processed. It is then a straightforward exercise to verify that the model's loose constraints are satisfied (line D in Figure 6-9) and that the observed step-by-step behavior conforms to the assumptions of the associated state transition diagram (line E in Figure 6-9). Once the loose constraints and the assumptions implicit in the state transition diagram are verified, the observed trajectory can be regarded as one of the trajectories associated with the LCD model. This linkage is represented by line G in Figure 6-9.

As noted previously, the mathematical relationships derived from an LCD model apply to all associated trajectories. Once it has been determined that the observed trajectory satisfies all necessary modeling assumptions, the mathematical relationship derived from the model should be valid for this trajectory as well. This is again illustrated by dotted line A.

The most common way to establish confidence in any model is represented by line J in Figure 6-9: simply compare the model's predictions with actual observed values. Of course, the success of any of these procedures for one or more intervals does not ensure that the model will continue to provide accurate results in the future. Nevertheless, successful verification of assumptions and validation of predictions can substantially increase the level of confidence in a model that has already passed the test of plausibility and has yielded satisfactory predictions in multiple thought experiments.

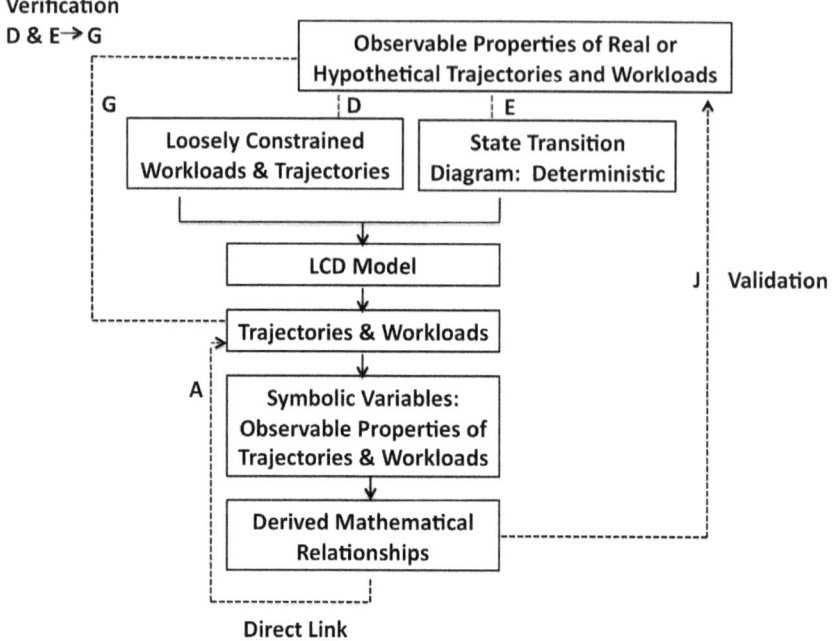

Figure 6-9. Verification and validation of LCD models

6.7.2 Modeling risk for stochastic models

Some of the strategies discussed in the preceding section are equally applicable to traditional stochastic models. In particular, the use of thought experiments to establish confidence in a stochastic model has exactly the same rationale.

On the other hand, the verification of modeling assumptions is significantly different in this case. The reasons lie in the subtle but crucial differences between Figures 6-9 and Figure 6-10. Note first that the state transition diagram, which represents the way a system reacts to various external events (e.g. arrival and departure of packets), is exactly the same in both cases. On the other hand, the workload driving this state transition diagram is not characterized by a set of directly verifiable loose constraints.   It is instead characterized in terms of the sampling premise. For the buffer overflow model, this premise leads to the assumption that each inter-arrival time and each packet transmission time can be regarded as a sample from an underlying probability distribution.

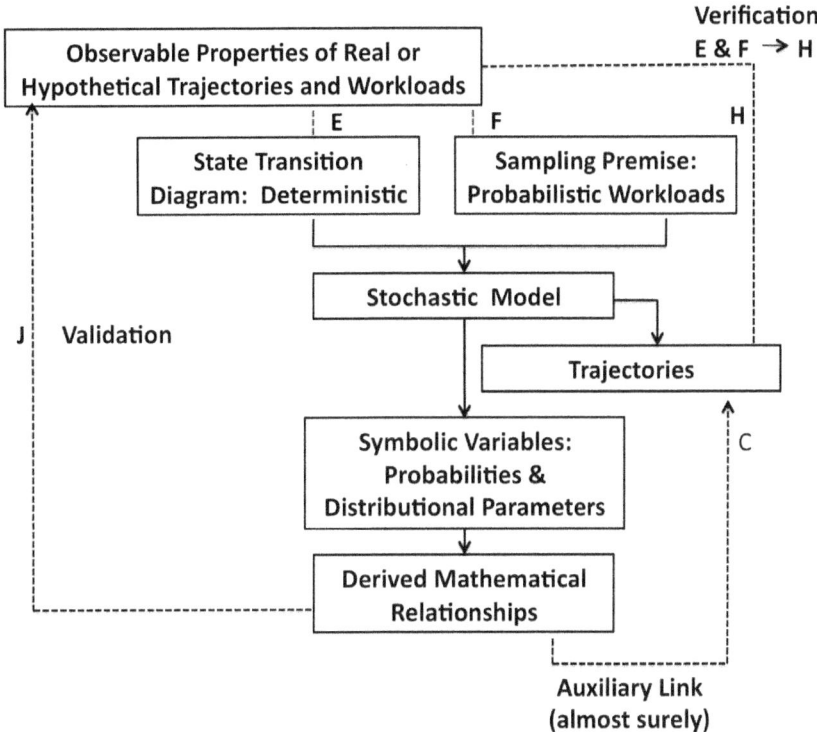

Figure 6-10. Verification and validation of stochastic models

The fundamental problem with the sampling premise is that it is not subject to direct verification (line F in Figure 6-10): there is simply no way to be certain that an observable quantity has been generated by sampling from an underlying probability distribution. The values presented in Table 6-2 and Table 6-3 underscore this point.   It is possible that these timestamps were generated by random sampling. It is also entirely possible that they were not.

If a large number of observations are available, statistical tests can be employed to determine the degree to which the entire set is consistent with the assumption that each observation has been drawn from the same probability distribution. Informally, these tests can strengthen the belief that the observables have been drawn from a probability distribution of a certain form. However, these statistical tests can never be conclusive because they deal with appearances rather than causality.

Since the validity of the sampling premise can never be established with certainty on the basis of observations alone, there is no way to be sure that the observed trajectory is associated with the stochastic model (line H). Despite these substantial reservations, it is still possible to validate the mathematical relationships derived from the stochastic model by comparing predicted values with observed values (line J). Success in validating these predictions builds confidence in the stochastic model, but is not sufficient to verify that the sampling premise is correct.

In summary, modeling risk for future experimental intervals exists for both LCD models and traditional stochastic models. It is never possible to be certain that the modeling assumptions made in either case will be satisfied in the future. However, for an interval that has actually been observed, the two cases are entirely different: the assumptions regarding loose constraints can be verified conclusively (line D in Figure 6-9), but assumptions based on the sampling premise (line F in Figure 6-10) cannot.

6.7.3 Forecasting risk

Equations (6-44) and (6-56) can both be used to predict the proportion of dropped packets when the capacity of the buffer is equal to three. Both equations are based on certain assumptions, and both equations entail modeling risk since the real systems to which they are applied may not satisfy these assumptions.

Suppose a practitioner who is responsible for managing the performance of a router in a real network is asked to predict the proportion of packets that will be dropped during a period of peak activity that is expected to occur six months in the future. The solution clearly depends on the rate at which packets arrive during that future period. This rate is represented by λ in equation (6-44) and by α in equation (6-56). Forecasting a future arrival rate is an entirely separate issue that is completely unrelated to the modeling problems that have been discussed thus far. The forecasting process lies instead within the realm of traditional statistics.

Some forecasting techniques deal directly with observable quantities without assuming that these observables have been drawn from underlying probability distributions: in other words, without invoking the sampling premise. Linear regression - based on a least squares fit - is perhaps the simplest example of this approach to forecasting.

Other forecasts are based on the premise that observables have been drawn from probability distributions that have a known form. This assumption creates a statistical model of the data. The goal of the analysis in these cases is to determine the most likely values of certain parameters that characterize the probability distributions upon which the statistical model is based.

Regardless of the approach that is employed, forecasting always involves uncertainty. This uncertainty is external to the models that have been discussed here, and is exactly the same for both observational stochastics and traditional stochastic modeling. The risk associated with these forecasts will be referred to as forecasting risk.

Note that forecasting risk does not arise in all modeling applications. For example, in the application discussed in Section 6.2.9, the goal is to determine the smallest buffer size required to maintain a 99.9% success rate (i.e., to keep the proportion of dropped packets below .001) when the external load is 200 packets per second and the packet processing rate is 250 packets per second. Many other design problems have exactly the same form: the desired external load is specified in the statement of the problem and need not be forecast, so the only risk is modeling risk.

Even when the external load is unknown, practitioners who apply models are not necessarily involved in the process of forecasting. For example, suppose the load on a computer system that supports the Human Resources department in some company depends on the number of employees working there. The job of the practitioner is to determine how fast the system must be to provide acceptable performance throughout the coming year.

The practitioner knows the costs and speeds of processors, disks and other components, but is unsure of the number of future employees since this number depends upon the rate of business expansion or contraction. Thus the overall load on the system is unknown. Senior management may have a good estimate of the number of future employees, but may not wish to share this estimate with the practitioner conducting the study.

The standard approach in such cases is to evaluate a range of alternatives and present the results to management for a final decision. Forecasting risk is shifted to senior managers who may have access to privileged information needed for reliable forecasts, and who may be more accustomed to making decisions in the face of uncertainty and incomplete information. This partitioning of responsibility among modelers, forecasters and decision makers is broadly applicable and quite common in practice.

6.7.4 Sampling risk

Sampling risk arises in all cases where observable values are regarded as samples that have been drawn from probability distributions.   The risk arises because it is never possible to guarantee that properties of a sample will be identical to corresponding properties of the distribution from which it is drawn. For example, there is no guarantee that the average inter-arrival time during some observation interval will be equal to the mean of the distribution from which inter-arrival times are drawn. In other words, it is impossible to guarantee that the observed value of λ will be equal to the value of the probabilistic parameter α.

If the two values differ, the observed value of $p(3)$ is likely to differ from the value predicted by equation (6-55) with α set to its theoretical value (line J in Figure 6-10). This represents a form of sampling risk. Note that sampling risk does not arise when results from observational stochastics such as equation (6-39) are used to predict the properties of actual trajectories (line J in Figure 6-9).

Even though sampling risk always exists in probabilistic models, this risk can be reduced by working with intervals that include large numbers of observations.   As discussed in Chapter 9, the Law of Large Numbers implies the chance that λ differs from α by any given value can be made arbitrarily small by using a trajectory with a suitably large number of inter-arrival times.

For purposes of this discussion, the main point is that sampling risk presents yet another concern for practitioners who employ stochastic models to obtain predictions about the observable behavior of real systems over finite intervals of time. Once again, this particular risk does not arise in observational stochastics: the symbolic variables of observational stochastics always correspond to quantities that are directly observable and measurable.

6.8 Observational stochastics: misconceptions

Because they are based on non-standard characterizations of uncertainty, observational stochastics and its predecessor operational analysis have sometimes been misunderstood by individuals with backgrounds in the classical theory of probability and stochastic processes. Several of these misconceptions are examined in this section.

6.8.1 Inflexible assumptions

In cases where assumptions such as matched endpoints or empirical independence are not satisfied, the accuracy of predictions that depend on these assumptions cannot be guaranteed. For example, in the buffer overflow model discussed in this chapter, the solution presented in equation (6-44) requires that the values of $\lambda(j)$ be empirically independent of j: all values of $\lambda(j)$ must be exactly equal to λ to guarantee the correctness of equation (6-44).

Critics have pointed out that assumptions such as these are unlikely to be satisfied exactly. They have gone on to assert that the assumptions employed in traditional stochastic models are more flexible and therefore superior. For example, the stochastic counterpart of the buffer overflow model includes the assumption of Poisson arrivals. This assumption, along with other assumptions, leads ultimately to a solution that has the same algebraic form as equation (6-44).

Critics note that a Monte Carlo simulation of the buffer overflow process will not necessarily generate values of $\lambda(j)$ that are empirically independent of j, even if the simulation faithfully implements the assumption of Poisson arrivals. This has led some critics to conclude that the assumption of Poisson arrivals is more flexible than the corresponding assumption of empirical independence.

Note that this apparent "flexibility" comes at a hidden cost: if the values of $\lambda(j)$ are not empirically independent of j, the solution expressed by equation (6-44) will not necessarily be correct, even in cases where a Monte Carlo simulation has faithfully incorporated the assumption of Poisson arrivals. In short, this type of flexibility is a questionable virtue – it is linked to the inability of probabilistic models to guarantee the accuracy of their predictions.

As one would expect, the correctness of equation (6-44) for a Monte Carlo simulation becomes increasingly likely as the length of the simulation grows. This is due in part to the fact that the values of $\lambda(j)$ are more likely to be empirically independent of j as the length of the simulation approaches infinity. In effect, the traditional stochastic assumption of Poisson arrivals leads to increasingly "inflexible" expectations regarding the values of $\lambda(j)$ as the length of the simulation increases.

6.8.2 Questionable utility

Observational stochastics deals with algebraic relationships among observable properties of trajectories and workloads. For example, equation (6-44) expresses $\bar{p}(N)$, the proportion of dropped packets when buffer size is equal to N, as a function of λ and μ. This result is valid for all trajectories with matched endpoints that satisfy the assumptions of empirical independence specified in Section 6.2.7.

In the view of some critics, such results are only meaningful in cases where a trajectory that satisfies the required assumptions actually exists so that λ and μ can be measured directly. In such cases, they argue that equation (6-44) is unnecessary: rather than measuring λ and μ and then using equation (6-44) to compute $\bar{p}(N)$, they argue it is more efficient to simply measure $\bar{p}(N)$ directly.

This criticism reflects a misunderstanding of the dual role that symbolic variables such as λ and μ play in mathematical modeling. In cases where a trajectory is available, the values of these variables can of course be measured directly, along with the actual value of $\bar{p}(N)$. These measured values can be used to validate the relationship specified by equation (6-44).

In applications that involve "what if" questions, actual trajectories are not available. Appropriate values of λ and μ must thus be estimated rather than measured. In cases where it is easier to estimate λ and μ than it is to estimate $\bar{p}(N)$ directly, the solution expressed by equation (6-44) is of genuine value. Considerations of this type provide the rationale for most mathematical models employed in science and engineering, including models that are developed within the framework of observational stochastics.

6.8.3 Misconstrued equivalences

Confusion also exists regarding the relationship between the stochastic assumption of Poisson arrivals in the traditional model of buffer overflow and the assumption that the values of $\lambda(j)$ are empirically independent of j. For models of this general type, it has been shown that the values of $\lambda(j)$ obtained from a stochastic trajectory will be empirically independent of j if and only if the arrival process is Poisson in the corresponding stochastic model.

As one might expect, this very interesting result only applies to the limiting case where the length of the trajectory approaches infinity. However, this type of equivalence does not imply that every trajectory encountered in the real world must have been generated by a Poisson process if the observed values of $\lambda(j)$ are empirically independent of j. A variety of other mechanisms, including some that incorporate both deterministic and probabilistic elements, are also capable of generating values of $\lambda(j)$ that are empirically independent of j.

The workloads in Tables 6-2 and Table 6-3 illustrate this point. As already noted, these workloads together with the state transition diagram in Figure 6-3 can generate trajectories with values of $\lambda(j)$ that are empirically independent of j.

Note that the values of $\lambda(j)$ continue to satisfy the assumption of empirical independence when these workloads are extended to include an indefinite number of repeating cycles. Also, each component used in the construction of such a repeating cycle can be selected arbitrarily from either Table 6-2 or Table 6-3. Workloads constructed in this manner have little in common with workloads generated by Poisson processes, except for the fact that the associated values of $\lambda(j)$ are empirically independent of j.

6.8.4 External uncertainty

Because random variables provide the foundation for traditional stochastic models, it has been argued that these models are especially well suited for representing the uncertainty that often arises when modeling results are applied to real world problems. It has also been argued that the deterministic nature of state transition diagrams leaves LCD models ill suited for the analysis of situations where uncertainty arises.

Chapters 1 and 3, along with earlier sections of this chapter, present a number of arguments that refute these claims. Note first that the uncertainty associated with forecasting the future values of symbolic variables is an external issue that affects LCD models and stochastic models in exactly the same manner: the uncertainty regarding the future value of λ in an LCD model is identical to the uncertainty regarding the future value of α in the corresponding stochastic model. The random variables incorporated into stochastic models have nothing to do with this form of uncertainty.

Uncertainty also exists regarding the validity of modeling assumptions. This second type of external uncertainty affects the degree of confidence that analysts have in a model: in particular, their confidence that a model accurately represents the factors that will determine the future behavior of the system being analyzed.

For the buffer overflow model, some practitioners may have confidence in the assumption of Poisson arrivals. Others may have confidence in the reasonableness of the assumption that future values of $\lambda(j)$ will be empirically independent of j. The future validity of either assumption is never completely certain, but the random variables incorporated into stochastic models are completely unrelated to this type of uncertainty. Thus stochastic models have no special advantage over LCD models for two of the most important forms of external uncertainty that are encountered in practice.

6.9 Sample path analysis

Sample path analysis (El-Taha and Stidham 1999) is an approach to the analysis of traditional stochastic models that is strikingly similar in spirit to observational stochastics. While observational stochastics deals with the properties of trajectories associated with LCD models, sample path analysis deals with the properties of trajectories associated with traditional stochastic models. This brief overview, which is intended primarily for students, compares the two approaches and also offers some insights into the relationship between models and their associated trajectories.

Figure 6-11 identifies the essential features of observational stochastics and sample path analysis. As in Figure 6-6, the left side of this figure refers to observational stochastics. Sample path analysis appears in place of traditional stochastic modeling on the right.

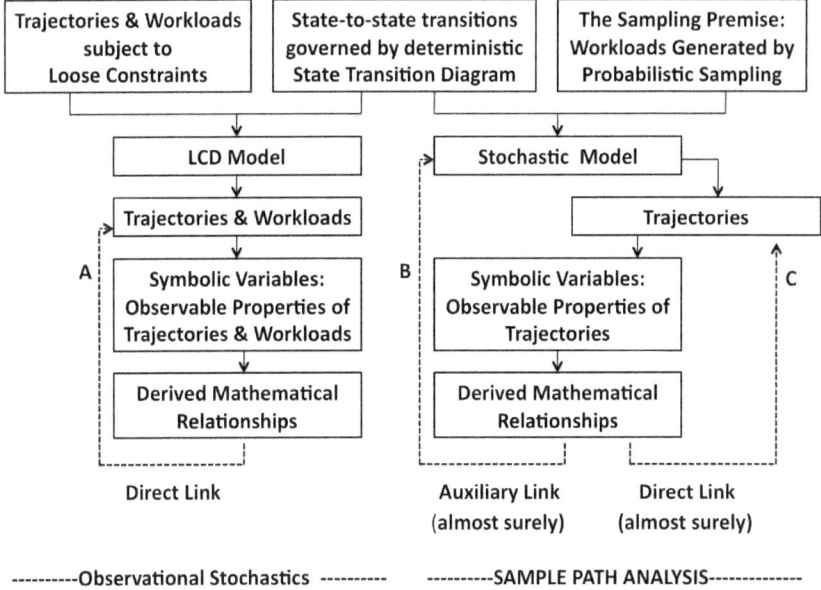

Figure 6-11. Sample path analysis

The most important similarity between observational stochastics and sample path analysis is that they both deal directly with the properties of trajectories. In other words, all symbolic variables and modeling assumptions that are employed in both approaches are defined in terms of observable and measurable properties of these trajectories. Note that this is quite different from traditional stochastic modeling, where symbolic variables such as α and β represent parameters of probability distributions associated with underlying stochastic processes.

The most important difference between these two approaches is the mechanism assumed to have generated the trajectory being analyzed. As already discussed, the trajectories studied in observational stochastics are generated when loosely constrained workloads are processed by finite state automata. In contrast, the trajectories studied in sample path analysis are simply abstract mathematical functions defined on the interval $[0,\infty)$. These functions are assumed to have certain limiting properties that are identical to the limiting properties of trajectories linked to associated stochastic processes. Informally, the trajectories studied in sample path analysis can be regarded as specific instances of the trajectories generated by Monte Carlo simulations that faithfully represent these associated stochastic processes.

The key insight behind sample path analysis is that trajectories of this type can be analyzed without referring to the parameters of the associated stochastic process and without invoking the mathematical assumptions required for the formal specification of that stochastic process. As in the case of observational stochastics, it is sufficient to deal exclusively with assumptions regarding observable properties of the trajectories themselves.

However, the ultimate objective in this case is to derive results that are applicable to the ensemble of trajectories linked to the associated stochastic process. To achieve this objective, sample path analysis deals primarily with results that are valid for trajectories of infinite length: technically, results that are valid in the limit as trajectory length approaches infinity.

Such results can only be derived if certain well defined limits are assumed to exist as trajectory length approaches infinity. Derivations carried out under assumptions of this type usually require delicate mathematical arguments involving upper and lower bounds on the limiting process (i.e., limits superior and limits inferior). Sample path analysis also requires carefully nuanced qualifications and reservations to deal with results that are almost surely, but not always, correct. A very brief introduction to the mathematical formalisms that have been developed to deal with such considerations is presented in Section 9.2.7.

None of these mathematical subtleties arise in observational stochastics because trajectories are always finite and are assumed to be generated by LCD models. Nevertheless, it is still interesting to know that the equations practitioners use to analyze finite length trajectories are also applicable to the limiting properties of trajectories that are associated with traditional stochastic processes, provided certain assumptions are satisfied. However, this information is not needed to justify the practical application of these equations within the context of observational stochastics.

CHAPTER 7

Extensions and Applications

7.1 Overview

The first few sections of this chapter extend the analysis in Chapter 6 to the general case where LCD models can include an arbitrary number of states and an arbitrarily connected state transition diagram. The development of these extensions closely follows the approach employed in Chapter 4; however, in this chapter it is assumed that time advances continuously rather than in discrete steps.

The main objective is to demonstrate that real world systems that are traditionally modeled as continuous time Markov processes can also be viewed through the lens of observational stochastics and modeled as continuous time LCD models. Even though the underlying assumptions are markedly different, the attained distribution derived from the LCD model will have exactly the same algebraic form as the steady state distribution derived from the corresponding Markov model. Thus, the two models yield exactly the same results for a large class of practical problems.

As in the case of Chapter 4, the mathematical notation required to deal with the general case adds complexity to the discussion. Readers interested in the applications of the analysis rather than the mathematical details of the general solution, may wish to review the following points and then proceed directly to Section 7.7. These seven points are direct analogs of their counterparts in Chapter 4.

1. The techniques used in Chapter 6 to analyze the buffer overflow problem can be extended to deal with continuous time LCD models where the number of states is equal to any value N and

where each state can be reached from every other state in a single step.

2. In this very general setting, the objective is to determine the proportion of time spent in each state over the course of the trajectory. These observable proportions are represented symbolically as $p(0)$, $p(1)$, $p(2)$ ... $p(N)$.

3. If the endpoints of a trajectory are matched, the desired values of $p(j)$ are given by the solution to a set of balance equations obtained, as usual, by setting the number of transitions into a state equal to the number of transitions out. These balance equations, which are specified in Theorem 7.1, are direct generalizations of equations (6-19) through (6-22).

4. If the endpoints of a trajectory are not matched, the solution to the balance equations still provides - in most cases - a reasonable approximation to the actual values of $p(j)$. In the limit as trajectory length increases, the error associated with this approximation usually approaches zero. If necessary, exact solutions for finite length trajectories with unmatched endpoints can be obtained through the procedure specified in Corollary 7.1.

5. The form of the general solution - with or without matched endpoints - is mathematically complex and of limited interest. Practitioners are typically concerned with solutions to balance equations that are associated with specific models such as the buffer overflow model in Chapter 6 and the various queuing models that will be discussed later in Chapter 7.

6. The t-loops introduced in Section 4.7.4 can be extended to represent continuous time processes. These extensions create richer structures with additional mathematical properties.

7. The general solution for the values of $p(0)$, $p(1)$, $p(2)$, ... $p(N)$ obtained using observational stochastics has exactly the same

algebraic form as the steady state distribution of the corresponding continuous time Markov process. However, the transition rate matrix that defines a traditional continuous time Markov process is comprised of instantaneous transition rates that must remain stationary (i.e., time-homogeneous) throughout the entire trajectory. In contrast, the global transition rate matrices employed in observational stochastics are comprised of average rates for entire trajectories. The detailed instant-by-instant forces that shape these trajectories have no impact on the analysis and need not be stationary. This again enables observational results to apply more broadly than the corresponding stochastic results.

7.2 Parameters for the general model

The buffer overflow example presented in Chapter 6 assumes that the capacity of the buffer is three packets. Suppose instead that the capacity of the buffer is N packets. This leads to an LCD model with N+1 states.

The buffer overflow example presented in Chapter 6 also assumes that at most one packet can arrive or depart at any instant. If this assumption is relaxed and bulk arrivals (i.e., multiple simultaneous arrivals) are permitted, it is possible for the number of packets in the buffer to increase from j to k in a single step for any value of k that is greater than j. Similarly, if bulk departures (i.e., multiple simultaneous service completions) are permitted, it is possible for the number of packets in the buffer to drop from j to k in a single step for any value of k that is less than j. Of course, j and k must both be greater than or equal to zero and less than or equal to N.

Assuming that bulk arrivals and bulk departures are both possible, the directly observable quantities $X(j,k)^{\#}$ and $T(j)^{\#}$ can be defined as follows.

$X(j,k)^{\#}$ = the number of times that state j is followed directly by
state k for j = 0, 1, 2 ...N and k = 0, 1,2 ... N

$T(j)^{\#}$ = total time spent in state for j = 0, 1, 2 ...N

Note that the definition of $X(j,k)^{\#}$ is identical to the corresponding
definition in Section 4.3.   Similarly, the definition of $T(j)^{\#}$ is
identical to the corresponding definition in Section 6.3.5.

The following symbolic variables can now be defined as functions
of these basic observables:

T = total length of the observation interval

$$= \sum_{j=0}^{N} T(j)^{\#} \qquad (7\text{-}1)$$

p(j) = proportion of time spent in state j

$$= T(j)^{\#} / T \qquad \text{for j=0, 1 ... N} \qquad (7\text{-}2)$$

x(j,k) = rate at which transitions from state j to state k occur
while in state j

$$= X(j,k)^{\#} / T(j)^{\#} \qquad (7\text{-}3)$$
$$\text{for j=0, 1 ... N} \qquad \text{and k=0, 1 ... N}$$

Note that the definitions of T and p(j) in equations (7-1) and (7-2)
are identical to the definitions of T and p(j) presented in Chapter 6.
In addition, the relationship between x(j,k) and X(j,k) is very
similar to the relationship between p(j) and P(j). As specified in
equation (4-6), X(j,k) is computed by dividing the number of
transitions from state j to state k by the total number of transitions
out of state j.   Similarly, x(j,k) is computed by dividing the

number of transitions from state j to state k by the total time spent in state j. Thus, X(j,k) is a dimensionless fraction while x(j,k) is a rate expressed as transitions per unit time.

In effect, the values of x(j,k) correspond to the parameters of a continuous time Markov process while the values of X(j,k) correspond to the parameters of a discrete time Markov process. As already noted, x(j,k) and X(j,k) represent observable quantities that are averaged over an entire observation interval. The corresponding Markovian parameters represent instantaneous stepwise values (in continuous time and discrete time models) that are assumed to remain invariant throughout the interval.

7.3 Analysis

The objective of the analysis is to express the values of p(j) as a function of the x(j,k). Note that this closely mirrors the analysis in Chapter 4, where the objective is to express the values of P(j) as a function of the X(j,k).

Theorem 7.1

Consider any continuous time trajectory generated by a formal LCD model. If the endpoints of this continuous time trajectory are matched, the values of p(j) defined in equation (7-2) are given by the normalized solution to the N+1 linear equations specified by equation (7-4).

$$p(j) \sum_{k=0}^{N} x(j,k) = \sum_{k=0}^{N} p(k) \times x(k,j) \quad \text{for } j=0, 1 \dots N \quad (7\text{-}4)$$

Proof:

Once again, the solution is based on a simple principle: for any trajectory with matched endpoints, the number of transitions into each state must be equal to the number of transitions out of that state. The argument that supports this conclusion has already been presented in the proof of Lemma 4.1.

Lemma 4.1, which is directly applicable to both the discrete time trajectories in Theorem 4.1 and the continuous time trajectories in Theorem 7.1, states that the values of $X(j,k)^{\#}$ for any trajectory with matched endpoints must satisfy equation (7-5).

$$\sum_{k=0}^{N} X(j,k)^{\#} = \sum_{k=0}^{N} X(k,j)^{\#} \qquad \text{for } j = 0, 1 \dots N \qquad (7\text{-}5)$$

Note that equation (7-5) is identical to equation (4-21).

The next step in the proof of Theorem 7.1 is to convert equation (7-5) into a relationship among the symbolic variables $p(j)$ and $x(j,k)$. As in the case of Theorem 4.1, this is a simple algebraic exercise. Equation (7-3) implies:

$$X(j,k)^{\#} = T(j)^{\#} \times x(j,k) \qquad (7\text{-}6)$$

Equation (7-2) implies that $T(j)^{\#}$ in equation (7-6) can be replaced by $T \times p(j)$.

$$X(j,k)^{\#} = T \times p(j) \times x(j,k) \qquad (7\text{-}7)$$

Replacing each occurrence of $X(j,k)^{\#}$ in equation (7-5) by $T \times p(j) \times x(j,k)$ yields

$$\sum_{k=0}^{N} T \times p(j) \times x(j,k) = \sum_{k=0}^{N} T \times p(k) \times x(k,j) \qquad (7\text{-}8)$$

$$\text{for } j = 0, 1 \dots N$$

Dividing both sides by T and simplifying the left hand side yields equation (7-4). The final step in the proof of Theorem 7.1 is to note that this system of N+1 linear equations, together with the usual normalizing constraint that requires the sum of all values of p(j) to be equal to 1, is sufficient to obtain a solution for each value of p(j). This completes the proof.

As in the case of Theorem 4.1, this proof guarantees the existence of a solution to the set of N+1 linear equations specified by equation (7-4) because values of p(j) that satisfy these equations can be extracted directly from the associated trajectory. Uniqueness of the normalized solution is assured because the matrix formed by the values of x(j,k) has a single irreducible subchain.

The general solution to the balance equations defined by equation (7-4) is too complex to be of practical interest. However, this solution can be simplified substantially for many models of genuine importance. The last few sections of this chapter will examine a number of these cases.

Note that equation (7-4) is similar in structure to equation (4-22), which represents the N+1 balance equations for a discrete time Markov chain. If x(j,k) is replaced by X(j,k) and p(j) is replaced by P(j), equation (7-4) becomes:

$$P(j) \sum_{k=0}^{N} X(j,k) = \sum_{k=0}^{N} P(k) \times X(k,j) \qquad (7\text{-}9)$$

$$\text{for } j = 0, 1 \dots N$$

In the discrete time analysis, the values of X(j,k) represent proportions rather than rates. As noted previously, the sum of the proportions leading out of each state must always be equal to 1. This constraint, which is expressed by equation (7-10), transforms equation (7-9) into equation (4-22).

$$\sum_{k=0}^{N} X(j,k) = 1 \qquad\qquad \text{for } j=0, 1 \dots N \qquad (7\text{-}10)$$

Note that the values of $x(j,k)$ for a continuous time trajectory do not necessarily satisfy a normalization constraint of the type specified by equation (7-10). This is the most important structural difference between the discrete time models in Chapter 4 and the continuous time models in Chapter 7.

7.4 Unmatched endpoints

As in the case of discrete time models, the assumption of matched endpoints is a technical convenience. Solutions can be derived whether or not this assumption is satisfied; however, the balance equations and their solutions are slightly simpler when endpoints are matched.

When the endpoints of a trajectory are not matched, it is still possible to use equation (7-8) to derive a solution for $p(0)$, $p(1)$, $p(2)$ ... $p(N)$. If the trajectory is at least moderately long, these computed values of $p(j)$ usually represent a good approximation of the actual values of $p(j)$ for the trajectory in question. This is the same conclusion derived in Section 4.7 for discrete time models.

When extending the arguments in Section 4.7.2 to continuous time models, the basic approach remains the same: convert the original trajectory into an adjusted trajectory with matched endpoints that has exactly the same values of $p(j)$. Then express the balance equations for this adjusted trajectory in terms of observable properties of the original trajectory (i.e., in terms of the values of $x(j,k)$ and other observable quantities).

Corollary 7.1

If the endpoints of a trajectory are unmatched, the values of p(j) are given by the normalized solution to the set of N+1 linear equations specified by equation (7-11).

$$p(j) \sum_{k=0}^{N} y(j,k) = \sum_{k=0}^{N} p(k) \times y(k,j) \qquad (7\text{-}11)$$

$$\text{for } j=0, 1 \dots N$$

In these equations, the value of y(j,k) represents the rate of transition from state j to state k while in state j – for an adjusted trajectory that is constructed by appending one additional state to the end of the original trajectory. The additional state has the same value as s(0), the initial state of the original trajectory, and the same timestamp as X, the special termination symbol. Thus the additional state appears between s(L) and X in the string of input symbols that define the adjusted trajectory.

Proof:

The endpoints of the adjusted trajectory are matched, so Theorem 7.1 is directly applicable to this trajectory. Moreover, the values of p(j) are identical in both the original trajectory and the modified trajectory since both trajectories have exactly the same length, and since no additional time is spent in state s(0) as a result of the modification. This completes the proof.

As in the case of Corollary 4.1, it is important to understand the relationship between the values of y(j,k) and the values of x(j,k). There is, in fact, only one minor difference. The adjusted trajectory has one additional transition from state s(L) to state s(0). If s(0) is once again set equal to a and s(L) is once again set equal to c, this extra transition will cause the value of y(c,a) to be slightly larger than the corresponding value of x(c,a).

The magnitude of the increase in x(c,a) depends on $T(c)^{\#}$, which has already been defined as the total amount of time spent in state c. Note that the value of $T(c)^{\#}$ is the same in both the original trajectory and the adjusted trajectory; however, the number of transitions from state c to state a in the adjusted trajectory is $X(c,a)^{\#}+1$. Thus

$$y(c,a) = x(c,a) + 1 / T(c)^{\#} \qquad\qquad (7\text{-}12)$$

If $T(c)^{\#}$ is at least moderately large, equation (7-12) implies that y(c,a) will be very close to x(c,a). As the length of the trajectory and the value of $T(c)^{\#}$ both increase, the maximum error introduced by assuming $y(c,a) = x(c,a)$ when determining the values of p(j) (i.e., by simply ignoring the fact that the endpoints of the trajectory may not be matched) becomes vanishingly small.

7.5 T-loops for continuous time models

The concept of a t-loop, which was introduced in Section 4.7.4, is directly applicable to continuous time trajectories. To construct a t-loop, characterize each transition point in a trajectory by two values: the identity of the state just entered, and the amount of time the system will remain in that state before the next transition occurs. Then connect the endpoint of the continuous time trajectory to the initial point. This creates a t-loop with exactly the same values of p(j) as the original trajectory.

If the endpoints of the trajectory are matched, the final state must be removed and the total amount of time in the final state at the end of the trajectory must be combined with the amount of time spent in the initial state at the beginning of the trajectory. The global transition rate matrix for this t-loop will then be exactly the same as the global transition rate matrix for the original trajectory.

If the endpoints are not matched, the final state must be retained and an additional transition from the final state to the initial state must be introduced at the point of attachment. This additional transition increases the transition rate from c to a. In this case equation (7-12) can be used to determine the new value of y(c,a). All other values of y(j,k) remain equal to the corresponding values of x(j,k).

Once again, the order in which states appear in a t-loop can be reversed to reflect a reversal in the flow of time. The amount of time spent in each state must be shifted by one position in this case. This will affect the values of x(j,k), but will leave the values of p(j) unchanged. The splicing operations described in Section 4.7.4 also have direct counterparts in this case.

7.6 Shaped simulation

The fact that balance equations (7-8) and (4-22) are sufficient to derive the correct values of p(j) and P(j) has an interesting implication for Monte Carlo simulations of continuous time and discrete time Markov chains. For a discrete time Markov chain, the elements of the transition matrix represent probabilities: the probability of transitioning from state j to state k is represented by the matrix element $\underline{X}(j,k)$. For a continuous time Markov chain, the elements of the transition matrix represent instantaneous rates: the instantaneous rate at which transitions from state j to state k take place is represented by the matrix element $\underline{x}(j,k)$. Note that neither $\underline{x}(j,k)$ nor $\underline{X}(j,k)$ represent directly observable quantities; they are instead parameters of stochastic processes.

Suppose a Monte Carlo simulation is being used to evaluate the steady state probability distribution for a discrete time or continuous time Markov chain. In such cases, the simulation is run for a period of time. In effect, each execution of a simulation program generates a trajectory of the type considered in observational stochastics. The observed values of p(j) or P(j) that are extracted from such trajectories are then assumed to represent the steady state distributions of the associated continuous time or discrete time Markov chains.

As noted previously, the justification for this conclusion is based on the Ergodic Theorem, which states that the observed values of p(j) or P(j) will, in the limit, almost surely be equal to the corresponding values of p(j) or P(j). This is the original rationale for Monte Carlo simulation.

The Ergodic Theorem does not provide any means for testing the output of a Monte Carlo simulation to ensure that the simulation has run long enough to produce an accurate result. Observational stochastics provides a new way to make this assessment: in addition to measuring the observed values of p(j), also measure the observed values of x(j,k). If the observed values of x(j,k) are all exactly equal to the underlying numerical values assigned to the x(j,k) at the start of the simulation, Theorem 7.1 guarantees that the simulation has generated values of p(j) that are exactly correct (subject to a minor adjustment for unmatched endpoints). The same principle applies to the values of P(j) generated by simulations of discrete time Markov chains.

In practice, the size of the transition matrix formed by the values of x(j,k) grows very quickly as the complexity of the Markov model increases. Whether or not this is a serious concern depends on the amount of storage that is available to the simulation program. In many cases of interest, the transition matrix is sparse and contains values that repeat according to a regular pattern. These two factors can reduce storage requirements dramatically.

Note that this new test of convergence for Monte Carlo simulations is only applicable to cases where the objective is to determine the steady state distribution of the underlying Markov chain (i.e., the values of $\underline{p}(j)$).   These cases are very common, and are the only ones considered here.   However, if analysts are interested in other properties of Markov chains that are not dependent upon the steady state distribution, there is no guarantee that the simulation will evaluate these *trans-distributional* quantities correctly.   This issue is discussed further in Section 7.9.3.

The new test of convergence also raises an interesting option for the designers of simulation programs.   Suppose a simulation has run for a period of time and generated a trajectory whose associated values of x(j,k) have been measured. Suppose further that these values do not match the corresponding underlying values of $\underline{x}(j,k)$.   In other words, suppose the simulation has not yet converged.   The traditional method for dealing with such situations is to continue running the simulation until a match arises.   Alternatively, it is possible to replace the random number generator that is driving the simulation by a non-random, goal oriented algorithm that attempts to extend the trajectory in a manner that causes the observed values of x(j,k) to converge quickly to the corresponding underlying values of $\underline{x}(j,k)$.

This replacement of a random number generator by a non-random goal oriented algorithm invalidates the probabilistic assumptions upon which the Monte Carlo simulation is based.   As a result, the Ergodic Theorem is no longer applicable to the trajectory that the simulation generates. Nevertheless, Theorems 7.1 and 4.1 imply that the values of p(j) or P(j) that are extracted from the trajectory will be exactly equal to the steady state distribution of the continuous time or discrete time Markov chain that is being simulated.

This alternative approach to simulation, which is referred to as shaped simulation (Buzen 2011), has been shown to reduce the execution time of simulations dramatically in a few simple, preliminary studies. Further investigation is required to assess the practical value of this approach. Note that the substantial speed advantage offered by shaped simulation is, in effect, balanced by the fact that these simulations can only be used to evaluate steady state distributions and their associated properties. Trans-distributional quantities cannot be evaluated with this approach.

7.7 Queuing systems

The balance equations represented by equation (7-4) are applicable to a wide variety of models. For many of these models, the matrix formed by the values of $x(j,k)$ has a simple structure that leads to equally simple balance equations and ultimately to closed form expressions for $p(j)$ that are easy to derive and understand. This is true for many queuing models.

Queuing phenomena arise in situations where customers arrive at servers, wait in queues, receive service, and then depart. Basic questions of interest include the lengths of queues, the utilizations of servers, the response times experienced by arriving customers, and the maximum processing rates for systems as a whole. Other questions involve the analysis of tradeoffs: for example, assessing the benefits gained by customers who receive preferential types of service, and contrasting these benefits with the extra delays experienced by customers who receive reduced levels of service.

Queuing theory is sometimes regarded as an advanced subject suitable only for students with strong backgrounds in calculus and probability theory. This is due to the fact that the solutions to many queuing problems are expressed in terms of the steady state distributions of continuous time stochastic processes. Observational stochastics provides an alternative approach to the analysis of queuing phenomena that is equally rigorous, but readily accessible to anyone with basic algebraic skills and an intuitive appreciation of the rationale behind the balance equations expressed by equation (7-4). Advanced training is not required to understand underlying principles and derive results of practical value. The simple queuing models examined in the next few sections illustrate these points.

7.8 Cyclic queues

The buffer overflow model discussed in Chapter 6 can be modified to create alternative models with important practical applications. One very simple modification is to assume that the supply of new packets shuts off completely once the buffer becomes full. Thus, overflow never occurs and packets are never lost. The state transition diagram for this case can be derived from Figure 6-3 by simply removing the transition arrow that originates from state 3, loops to the right, and returns to state 3. The modified state transition diagram is shown in Figure7-1.

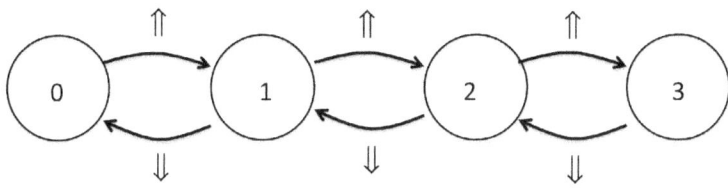

Figure 7-1. State transition diagram for a cyclic queue

Figure 7-2 provides an alternative representation of the cyclic queuing system characterized by Figure 7-1. The rectangles in Figure 7-2 represent the location of queues (i.e., buffers) and the circles represent the location of servers. Customers circulate counter-clockwise around this closed loop, visiting each server once per cycle. If a customer arrives at a server while it is busy serving another request, the customer simply waits in the associated queue until the server becomes available. If all customers are at server A, no additional arrivals can occur. The arrival process simply shuts down.

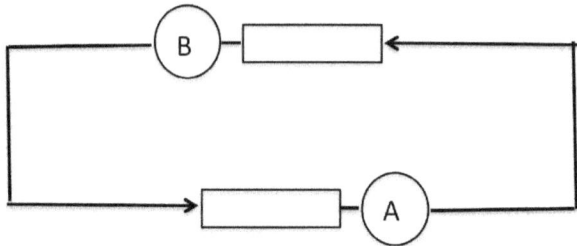

Figure 7-2. Cyclic queue

Diagrams of this type will be used in the remaining sections of this chapter. They provide a convenient way to describe the paths that customers follow as they flow through a queuing system. In this case, the number of customers at server A represents the state of this cyclic queuing system. Since there are three customers circulating around this loop, the number of customers at server A is either 0, 1, 2 or 3. These values correspond to the states depicted in Figure 7-1.

7.8.1 Analysis of a cyclic queue

Suppose the objective of an analysis is to determine the values of p(j): that is, the proportion of time that the number of customers at server A is equal to 0, 1, 2 or 3. These proportions clearly depend on the amount of service customers require during each visit to servers A and B.

To apply the buffer overflow analysis to this problem, simply replace the concept of packets arriving at a buffer by the concept of customers arriving at the queue associated with server A. The directly observable quantities defined in Sections 6.3.5 and 6.3.6 can then be associated with the cyclic queue in Figure 7-2. In particular, the definitions in equations (6-8) and (6-9) can be updated as follows:

$\lambda(j)$ = arrival rate of customers at server A during those times when the queue currently contains j customers

$\mu(j)$ = rate at which customer requests are processed by server A during those times when the queue currently contains j customers

Replacing the input symbols in Figure 7-1 with the rates at which the corresponding transitions occur yields Figure 7-3. The local balance equations for this model, which can be identified directly from this diagram, are given by equations (7-13) through (7-15). These equations are identical to the local balance equations for the buffer overflow model (i.e., equations (6-23) through (6-25)).

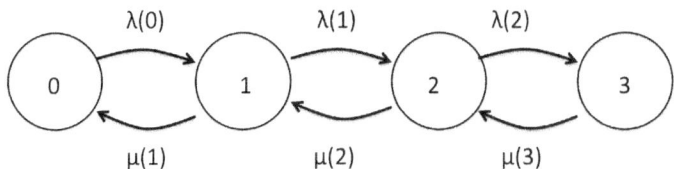

Figure 7-3. Input symbols replaced by transition rates

$$\lambda(0)\times p(0)=\mu(1)\times p(1) \tag{7-13}$$

$$\lambda(1)\times p(1)=\mu(2)\times p(2) \tag{7-14}$$

$$\lambda(2)\times p(2)=\mu(3)\times p(3) \tag{7-15}$$

Equations (7-16) through (7-19) present the normalized solution to equations (7-13) through (7-15).

$$p(0)=\cfrac{1}{1+\cfrac{\lambda(0)}{\mu(1)}+\cfrac{\lambda(0)\times\lambda(1)}{\mu(1)\times\mu(2)}+\cfrac{\lambda(0)\times\lambda(1)\times\lambda(2)}{\mu(1)\times\mu(2)\times\mu(3)}} \tag{7-16}$$

$$p(1) = \frac{\lambda(0)}{\mu(1)} \times p(0) \qquad\qquad (7\text{-}17)$$

$$p(2) = \frac{\lambda(0) \times \lambda(1)}{\mu(1) \times \mu(2)} \times p(0) \qquad\qquad (7\text{-}18)$$

$$p(3) = \frac{\lambda(0) \times \lambda(1) \times \lambda(2)}{\mu(1) \times \mu(2) \times \mu(3)} \times p(0) \qquad\qquad (7\text{-}19)$$

These equations are of course identical to equations (6-29) through (6-32). Despite these similarities, there is an obvious difference between the two models: in the buffer overflow model, packets continue to arrive while the buffer is full. In other words, $\lambda(3)$ has a positive value. However, the value of $\lambda(3)$ does not appear in equations (6-29) through (6-32), so it has no effect on the solution for the values of p(j).

The impact of $\lambda(3)$ becomes apparent when considering the value of λ, the overall rate at which customers arrive at server A. This quantity is equal to the total number of arrivals divided by the total length of the observation interval:

$$\lambda = \frac{A(0)^{\#} + A(1)^{\#} + A(2)^{\#} + A(3)^{\#}}{T(0)^{\#} + T(1)^{\#} + T(2)^{\#} + T(3)^{\#}} \qquad\qquad (7\text{-}20)$$

If the arrival rate $\lambda(j)$ is empirically independent of queue length in the buffer overflow model, each value of $\lambda(j)$ will be equal to λ. However, for the cyclic queue, empirical independence implies that $\lambda(0)$, $\lambda(1)$ and $\lambda(2)$ are all the same, but $\lambda(3)$ is equal to zero. Define $\hat{\lambda}$ (pronounced "lambda hat") as the overall rate at which customers arrive at server A during those times when arrivals are possible. Since no arrivals are possible when all three customers are at server A, $\hat{\lambda}$ and λ are not the same. Equation (7-21) specifies the relationship between these two arrival rates.

$$\lambda = \hat{\lambda} \times \left[p(0) + p(1) + p(2) \right] \qquad (7\text{-}21)$$

Note that the overall arrival rate λ is less than $\hat{\lambda}$ because the overall rate includes times when the queue is full and the arrival rate at server A drops to zero. This is important for the discussion in Section 7.9 because Little's Law, a fundamental result in queuing theory, depends on λ rather than $\hat{\lambda}$.

7.8.2 Limit as N approaches infinity

If $\lambda(0) = \lambda(1) = \lambda(2) = \hat{\lambda}$ and $\mu(1) = \mu(2) = \mu(3) = \mu$, equations (7-16) – (7-19) reduce to equations (7-22) – (7-25).

$$p(0) = \frac{1}{\displaystyle\sum_{j=0}^{3} \left(\hat{\lambda}/\mu \right)^{j}} \qquad (7\text{-}22)$$

$$p(1) = \frac{\hat{\lambda}/\mu}{\displaystyle\sum_{j=0}^{3} \left(\hat{\lambda}/\mu \right)^{j}} \qquad (7\text{-}23)$$

$$p(2) = \frac{\left(\hat{\lambda}/\mu \right)^{2}}{\displaystyle\sum_{j=0}^{3} \left(\hat{\lambda}/\mu \right)^{j}} \qquad (7\text{-}24)$$

$$p(3) = \frac{(\hat{\lambda}/\mu)^3}{\displaystyle\sum_{j=0}^{3}(\hat{\lambda}/\mu)^j} \tag{7-25}$$

Equations (7-22) through (7-25) can be simplified further by allowing the number of circulating customers in Figure 7-1 to become very large. As this number increases without limit, it is easy to show that the denominators in these equations converge to $1 - \lambda/\mu$, provided $\lambda > \mu$ (i.e., provided that rate at which server A is able to process customer requests is greater than the rate at which these requests arrive). This simplification yields the following limiting solution:

$$p(j) = \frac{(\lambda/\mu)^j}{1 - \lambda/\mu} \qquad \text{for } j = 0, 1, 2 \ldots \tag{7-26}$$

Equation (7-26) corresponds to the steady state distribution of a classical M/M/1 queue. This is often the first result derived in a traditional course on queuing theory. Despite its prominence, it is important to remember that equation (7-26) represents the exact solution to an idealized problem that can never be encountered in practice. In the real world, there is always a finite upper bound N that queue length can never exceed. Values of $p(j)$ beyond this limit must be zero rather than the values $p(j)$ that are given by equation (7-26).

In practice, the error introduced by employing equation (7-26) and its variants is often minor. As part of a systematic study of this issue, Buzen and Goldberg (1974) show that predictions of average response time based on equation (7-26) have an accuracy of 95% or better when N is at least 38 and λ/μ is less than one half. Achieving the same level of accuracy when $\lambda/\mu = 0.7$ requires a value of N that is at least 142.

As λ/μ approaches 1.0, the value of N required to achieve acceptable accuracy increases substantially.

Figure 7-4 provides a way to conceptualize the idealized queuing system that corresponds to equation (7-26). Once again, this system is comprised of two servers: A and B. However, these servers are not arranged to form a closed loop. Instead, the customers are first processed by server B, then move on to server A, and finally exit from the system. The system is initialized with an infinite number of customers waiting in the queue for server B. The analysis is focused exclusively on the length of the queue at server A, so the actual length of the queue at server B is irrelevant. All that matters is that server B remains active throughout the interval.

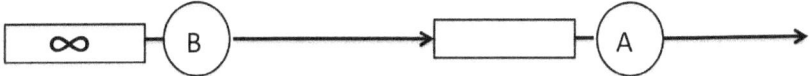

Figure 7-4. A queuing model with infinite source arrivals

To represent the system in Figure 7-4 as a traditional stochastic process, simply assume that the amount of time required to process customer requests at servers A and B are drawn, at random, from two probability distributions. If the distribution used to determine service times at server B is exponential with mean $1/\lambda$, and if the distribution used to determine service times at server A is exponential with mean $1/\mu$, the resulting stochastic process is classified as an M/M/1 queue (where the first M indicates that inter-arrival times are Markovian, the second M indicates the service times at server A are also Markovian, and the integer 1 indicates that only one server is available to process customers waiting in the queue at server A). M/M/1 queues and their variants provide the foundation for elementary queuing theory.

From a practitioner's perspective, the infinite source model in Figure 7-4 offers an important advantage: it has only two parameters, λ and μ. In contrast, the cyclic queue in Figure 7-2 requires three parameters: $\hat{\lambda}$, μ, and N, the number of circulating customers in the closed loop. Note that λ and μ can be measured by instruments positioned locally at server A. On the other hand, $\hat{\lambda}$ is most easily measured at server B, and the direct measurement of N requires simultaneous access to the queue lengths at both servers. This issue will be revisited in Section 7.10.

7.9 Queue length and response time

The analyses in Section 7.8 focused on the number of customers at server A: specifically, the proportion of time for which this number is equal to 0, 1, 2 and so on. In many queuing models, analysts are also concerned with response time: the total amount of time required to complete a customer's request, including time spent waiting in a queue plus time spent actually receiving service.

7.9.1 Little's Law

Intuitively, it is reasonable to expect that average response time will increase as average queue length increases. The relationship between these two quantities is governed by a result known as Little's Law (Little 1961). The intuitive insight that lies behind this result is easy to understand. Consider any server that processes requests generated by arriving customers. As usual, let λ represent the rate at which customers arrive at this server.

Now define two additional quantities:

L = average number of customers at the server
(waiting for or receiving service)

R = average response time for arriving customers

It is clear that L and R represent observable quantities that can be measured over intervals of time. The precise definitions of L and R will be presented shortly, but are not required for this informal discussion.

Suppose that an observer examines the server at some arbitrary point in time. Since R is the average response time, those customers who arrived within the past R seconds will, on the average, still be present (i.e., waiting for or receiving service). On the other hand, those customers who arrived more than R seconds ago will, on the average, be gone. Thus L, the average number of customers at the server, will generally be equal to the number of customers who arrived in the past R seconds.

Since λ is the rate at which customers arrive at this server, the number of customers expected to arrive in the past R seconds is simply λR. Thus

$$L = \lambda R \qquad (7\text{-}27)$$

Equation (7-27) represents the essence of Little's Law. Although this result is broadly applicable, its most important uses are in cases where L, the average number of customers at the server, can be derived from a queuing model (e.g., the model in Figure 7-2 or Figure 7-3). Little's Law can then be used to determine the average response time for customers in the system being modeled.

For the queuing models discussed in the preceding section, p(j) is the proportion of time that the number of customers at server A is equal to j. Thus L, the average number of customers at server A over the entire interval, is given by equation (7-28). This equation provides the formal definition of L.

$$L = \sum_{j=0}^{N} j \times p(j) \qquad (7\text{-}28)$$

For infinite source queues of the type illustrated in Figure 7-3, the average arrival rate λ that appears in equation (7-27) is a basic parameter of the queuing model. However, the arrival rates for cyclic queues of the type illustrated in Figure 7-2 are expressed in terms of $\lambda(j)$. These quantities are related to λ by equation (7-29).

$$\lambda = \sum_{j=0}^{N-1} \lambda(j) \times p(j) \qquad (7\text{-}29)$$

In the special case where the values of $\lambda(j)$ are equal to $\hat{\lambda}$ for $j = 0$, 1, 2 ... N-1, and where $\lambda(N)$ is equal to zero, equation (7-28) can be simplified as follows:

$$\lambda = \hat{\lambda} \sum_{j=0}^{N-1} p(j)$$

$$= \hat{\lambda}\left[1 - p(N)\right] \qquad (7\text{-}30)$$

As N becomes large, p(N) approaches zero provided λ/μ is less than 1. Thus $\hat{\lambda}$ approaches λ in these cases.

7.9.2 Response time for an M/M/1 queue

The values of p(j) for a general cyclic queue are given by equations (7-16) through (7-19). In the special case where $\lambda(0)$, $\lambda(1)$ and $\lambda(2)$ are all equal to a common value $\hat{\lambda}$, and where $\mu(1)$, $\mu(2)$ and $\mu(3)$ are all equal to a common value μ, the simplified solution given by equations (7-22) through (7-25) is sufficient.

No special insights can be gained by applying equation (7-28) to either of these cases. However, it is interesting to consider the value of L in the limiting case where N approaches infinity and the values of $\lambda(j)$ and $\mu(j)$ are empirically independent of j. This corresponds to the M/M/1 infinite source model in Figure 7-3. In this case, equations (7-26) and (7-28) imply:

$$L = \frac{\lambda/\mu}{1-\lambda/\mu} \qquad (7\text{-}31)$$

Combining equations (7-31) and (7-27) then yields what is arguably the most important result in basic queuing theory:

$$R = \frac{1/\mu}{1-\lambda/\mu} \qquad (7\text{-}32)$$

The implications of equation (7-32) become apparent when μ and λ/μ are replaced by alternative variables. As defined by equation (6-7), μ is the overall rate at which customer requests are processed by server A. This rate is equal to C/B, where C is the total number of completed requests and B is the total amount of time the server is busy. The reciprocal of μ, which is equal to B/C, is the average amount of processing time per completed request (i.e., average service time). As specified in equation (6-6), this quantity is represented by the symbolic variable S. Thus, $1/\mu$ can be replaced by S in the numerator of equation (7-32).

Now consider the term λ/μ in the denominator of equation (7-32). Since λ is defined as A/T by equation (6-5),

$$\lambda/\mu = \frac{A}{T} \times \frac{B}{C} \qquad (7\text{-}33)$$

If the endpoints of a trajectory are matched, the total number of customers who arrive at server A will be equal to the total number of completed requests at server A. In other words, A will be equal to C, which implies that λ/μ is equal to B/T.

Since B is the total amount of time server A is busy, the utilization of server A during the entire interval is B/T as indicated in equation (7-34)

U = utilization of server A

$$= B / T \qquad\qquad (7\text{-}34)$$

Replacing λ/μ by U and $1/\mu$ by S transforms equation (7-32) into the more convenient form that is shown below in equation (7-35): average response time R is equal to average service time S divided by (1-U).

$$R = \frac{S}{1-U} \qquad\qquad (7\text{-}35)$$

It is clear from equation (7-35) that the average response time for arriving customers is approximately equal to average service time S when utilization is low. This makes sense since queuing delays should be very small in such cases. When utilization is in the neighborhood of 50% (i.e., $U = 0.50$), average response time is approximately equal to twice average service time. In other words, for every second of service a customer receives, the customer also experiences one second of queuing delay.

Beyond 75%, average response time increases rapidly as a function of U. When utilization is 75%, response time is four times service time. It increases to five times service time at a utilization of 80% and to ten times service time at 90%. This classic queuing behavior is illustrated in Figure 7-5.

7.9.3 Trans-distributional nature of response time

Little's Law makes it possible to express average response time R as a function of λ and the values of p(j). While average response time is important, analysts are often interested in additional information such as the proportion of customers whose response time exceeds some desired threshold. In some cases, the complete distribution of response time is also of interest.

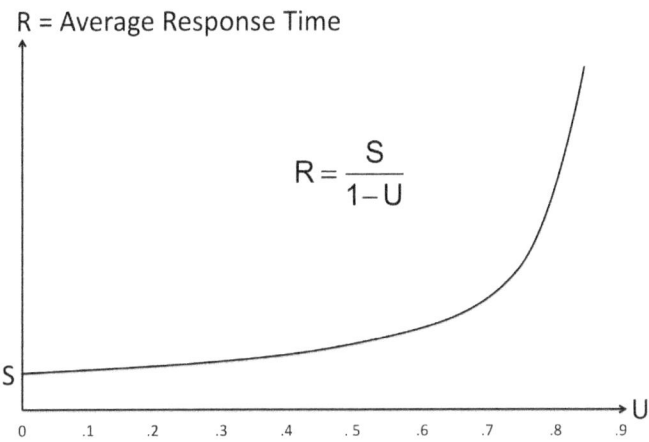

Figure 7-5. Average response time as a function of utilization

This raises an obvious question: is it possible to express the distribution of response time as a function of the values of p(j) and the parameters $\lambda(j)$ and $\mu(j)$? The answer is no. To understand why, consider the trivially simple case of a cyclic queue with only one circulating customer. This model has only two parameters: $\lambda(0)$ and $\mu(1)$. As usual, $1/\lambda(0)$ is the average time between arrivals, and $1/\mu(1)$ is the average service time per customer request.

In this trivially simple case, there are never any queuing delays. Thus, p(0) is equal to $\mu(1)/[\lambda(0)+\mu(1)]$ and p(1) is equal to $\lambda(0)/[\lambda(0)+\mu(1)]$. Moreover, each customer's response time is always exactly equal to that customer's service time. It follows immediately that average response time R is equal to average service time $1/\mu(1)$. However, since no other information has been provided regarding the service time distribution, nothing else can be deduced about the response time distribution. This conclusion extends to the general case where the number of circulating customers is equal to N and where response time is equal to service time plus a non-zero queuing delay.

The more powerful assumptions employed in traditional stochastic queuing models do in fact make it possible to derive expressions for response time distributions. However, these analyses are considerably more complex and dependent upon assumptions that go well beyond the observational assumptions required to derive queue length distributions.

From a practitioner's perspective, reliance upon additional assumptions introduces additional risk. Thus queuing equations that predict response time distributions are inherently riskier than equations that predict queue length distributions, average queue length, and – via Little's Law – average response time. The fact that a model provides accurate predictions of these lower-risk quantities does not insure that predictions of response time distributions or response time percentiles will be accurate.

Using the terminology introduced earlier, all properties of response time – with the exception of average response time – are trans-distributional quantities. They cannot be expressed as functions of the queue length distribution (i.e., $p(0)$, $p(1)$, $p(2)$ ...) unless additional assumptions are introduced. This implies, among other things, that the shaped simulation technique described in Section 7.6 cannot be used to compute response time distributions even though this technique is capable of accurately computing queue length distributions and average response time.

One other practical implication of this discussion concerns the use of service level agreements (SLAs) that are common in computer/communications environments. For example, an SLA might specify that 90% of all transactions must be completed in less than two seconds. SLAs based on response time percentiles are usually regarded as preferable to SLAs based on average response time (e.g., requiring average response time to be less than one second). However, such SLAs are also riskier because they are sensitive to factors that are immaterial when calculating average response time.

A more robust approach to constructing SLAs is to state them in terms of queue length distributions. For example, an SLA might require that 90% of arriving transactions must encounter a queue whose length is less than ten. This SLA depends entirely on distributional quantities: in particular, the values of $\vec{p}(j)$. By equation (6-42), these values are simple functions of $p(j)$ and are thus distributional quantities. This can be seen in equation (6-36). The values of $\lambda(j)$ in equation (7-36) are basic parameters of the model; λ is specified by equation (7-29).

$$\vec{p}(j) = \frac{\lambda(j)}{\lambda} \times p(j) \qquad \text{for } j = 0, 1, 2 \ldots \qquad (7\text{-}36)$$

7.9.4 Little's Law: sketch of proof

Within the framework of observational stochastics, the formal proof of Little's Law is based on an alternative method for characterizing the average number of customers at server A. The standard definition of L is given by equation (7-28). It is valid for all trajectories, whether or not their endpoints are matched.

To construct an equivalent alternative definition, suppose as usual that a server is observed over an interval of T seconds and that A equals the number of customers who arrive during the interval. Let A^+ represent the value of A plus the number of customers at the server at the start of the interval. Thus A^+ is the total number of customers who spend at least some time at the server during the interval.

Now consider the idea of residency time. For each of the A^+ customers identified in the preceding paragraph, let r(j) equal the total number of seconds spent by that customer at the server (waiting in the queue or receiving service). The value of r(j) represents the residency time for customer j.

The next step is to consider the sum of the values of r(j) over all A^+ customers. This sum represents the total accumulated residency time during the entire interval. As shown in equation (7-37), dividing this total by the length of the interval T yields the average number of customers in residence during the interval (i.e., the average queue length). This subtle but crucial point is the key to the proof of Little's law. The rest of the proof is simple algebra.

A^+ = total number of customers who spend at least some time at the server

r(j) = residency time at the server for customer j (j = 0, 1, ... A^+)

$$L = \frac{1}{T} \sum_{j=1}^{A^+} r(j) \tag{7-37}$$

The intuitive justification for equation (7-37) is clear. For a formal derivation of this equation, see (Denning and Buzen 1978) or standard contemporary references such as (Menascé and Almeida 2001), (Bolch, Greiner, de Meer and Trivedi 2006), (Gelenbe and Mitrani 2010), (Kobayashi , Mark and Turin 2012) and (Harchol-Balter 2013).

The next step is to replace 1/T by (1/A)×(A/T) and to observe that A/T is equal to λ. This transforms equation (7-37) into equation (7-38).

$$L = \lambda \times \frac{1}{A} \sum_{j=1}^{A^+} r(j) \qquad (7\text{-}38)$$

The final step is to set W equal to the total accumulated residency time divided by A, the total number of arrivals. As discussed in Section 7.9.5, W represents an imperfect measure of response time that is distorted by end effects.

W = average residency time per arrival

$$= \frac{1}{A} \sum_{j=1}^{A^+} r(j) \qquad (7\text{-}39)$$

Combining this definition of W with the equation (7-38) yields the classic form of Little's Law given by equation (7-40)

$$L = \lambda W \qquad (7\text{-}40)$$

Note that the classic version of Little's Law is expressed in terms of average residency time W rather than average response time R. The next section explores the relationship between these two quantities.

7.9.5 End effects

Response time is an especially easy concept to understand. It is simply the time between a customer's arrival and departure. However, when the operation of a server is observed during a finite interval of time, it becomes necessary to consider the response times of customers who are already at the server when the interval begins, as well as customers who remain at the server when the interval ends.   The response times of these customers cannot be determined since matched sets of arrival and departure times are unavailable.

The term *end effects* refers to the impact that these customers have on the observed values that are used to construct queuing models. In practice, response times are typically measured at the instant a customer completes service and is ready to leave the server. Thus measured response times for customers who are already present when the interval begins can be unexpectedly large. In addition, customers who have not yet completed their service when the interval ends contribute to measured server utilization and queue length, but have no impact at all on measured response times.

One way to address this problem is to replace the concept of response time by the concept of residency time. This is the motivation for the symbolic variable W defined in equation (7-39). Note that residency time only refers to time accumulated during the interval itself. In effect, equation (7-39) is most useful in cases where the amount of residency time accumulated prior to the start of the observation interval (by customers who are present when the interval begins) is approximately equal to the amount of residency time yet to be accumulated by customers still present when the interval ends.

If the observation interval begins and ends at points where the queue is empty (i.e., no customers present at the server), end effects are eliminated and residency time is exactly equal to response time for all customers. This assumption is implicit in the simplified version of Little's Law that appears in equation (7-27).

7.10 Finite source arrivals: the machine repairman model

The machine repairman problem is one of the classic problems in queuing theory. Imagine a factory that utilizes a number of machines to produce products of various types. Each machine operates independently. After a machine has been running for a while, it breaks down and needs to be repaired. There is a single repairman assigned to fixing these broken machines. The problem is to determine the productivity rate of the factory (i.e., the average number of machines in service) as a function of the total number of machines, the mean time between failure, and the mean time to repair a broken machine.

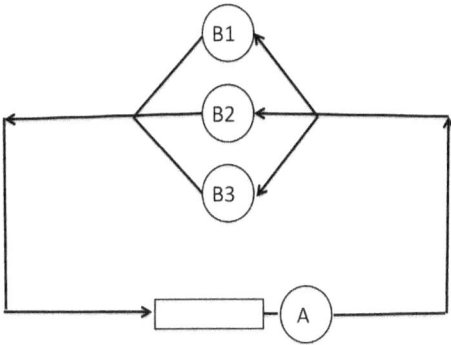

Figure 7-6. A queuing model with finite source arrivals

Figure 7-6 illustrates a queuing model that incorporates the essential elements of this problem. To maintain consistency with the buffer overflow analysis in Chapter 6 and the cyclic queue analysis in the preceding section, Figure 7-6 depicts the simple case where there are only three machines in the factory. However, the analysis extends directly to cases where the number of machines is equal to any integer N.

The analysis begins with a concept that may at first seem surprising: even though the machines are stationary and the repairman moves around the factory to make repairs, the customers in this queuing model are the machines, and the roaming repairman is the server. It is convenient to think of each machine as being represented by a token that moves around the closed loop in Figure 7-6. When a token is at server B1, B2 or B3, the corresponding machine is operating properly. When that token is at server A, the corresponding machine is broken and is either being repaired or waiting in the queue for service.

Note that the state transition diagram in Figure 7-1 is once again applicable to this model. Thus the balance equations presented in equations (7-13) through (7-15) are also applicable. The only differences between the solutions to the models in Figures 7-2 and 7-6 are the assumptions made regarding the values of $\mu(j)$ and $\lambda(j)$.

Recall that μ represents the rate at which the server completes customer requests. Suppose it takes an average of 15 minutes to repair a machine. Then the overall rate at which the repairman (i.e., server A) can operate is 4 repairs per hour or 32 repairs per eight hour shift. This is, of course, an overall rate. For this simple analysis, assume the rate at which repairs are made is empirically independent of the length of the queue. In other words, assume that the values of $\mu(1)$, $\mu(2)$ and $\mu(3)$ are all equal to 32, which is the value of μ in this specific case.

The values of $\lambda(j)$ must be handled differently. Begin by considering M, the mean time between failure. For this example, suppose that machines break down after an average of four hours of actual service. Thus M is equal to one half of an eight hour shift (i.e., $M = \frac{1}{2}$).

Now consider the value of $\lambda(2)$, which is the breakdown rate during those times when there are two machines at server A: one being repaired, and one waiting for repair. There is exactly one machine in service at such times. Since one machine generates an average of two breakdowns per eight hour shift, it is reasonable to expect that $\lambda(2)$ will be equal to 2 in this particular case and will be equal to 1/M in general.

Now consider the value of $\lambda(1)$, which is the breakdown rate during those times when there is only one broken machine at server A. There are exactly two machines in service at such times. Each generates an average of two breakdowns per eight hour shift.

Thus it is reasonable to expect that $\lambda(1)$ will be equal to 4 in this particular case and 2/M in general.

Similar reasoning applies to $\lambda(0)$, which is the breakdown rate during times when all three machines are in service. In general, it is reasonable to expect that $\lambda(0)$ will be equal to 6 in this particular case and 3/M in general.

7.10.1 Online and off-line behavior

These intuitive arguments regarding the values of $\lambda(j)$ can be formalized through an assumption that is applicable to many different queuing models.   This is the assumption that online behavior is equal to off-line behavior (Denning and Buzen 1978): in other words, the online behavior of a group of servers within a network is equal to the behavior of that same group when it is detached from the rest of the network (i.e., taken off-line) and then studied in a systematic fashion.

In this example, servers B1, B2 and B3 are taken off-line by detaching them from server A (i.e., the rest of the network) and then connecting their outputs directly back to their inputs. This is shown in Figure 7-7.

The next step is to determine the throughput rate of the off-line subsystem when the number of circulating tokens is equal to 1, 2 and 3. These throughput rates which are denoted by $g^*(1)$, $g^*(2)$ and $g^*(3)$, must be expressed as a function of M, the mean time between failure.

The value of $g^*(1)$ is trivial to determine.   In this case one token cycles endlessly around the loop shown in the off-line model at the bottom of Figure 7-7. The average time per cycle is the average service time at server B1. This average service time is equal to M. Thus the throughput rate $g^*(1)$ is equal to 1/M cycles per unit time.

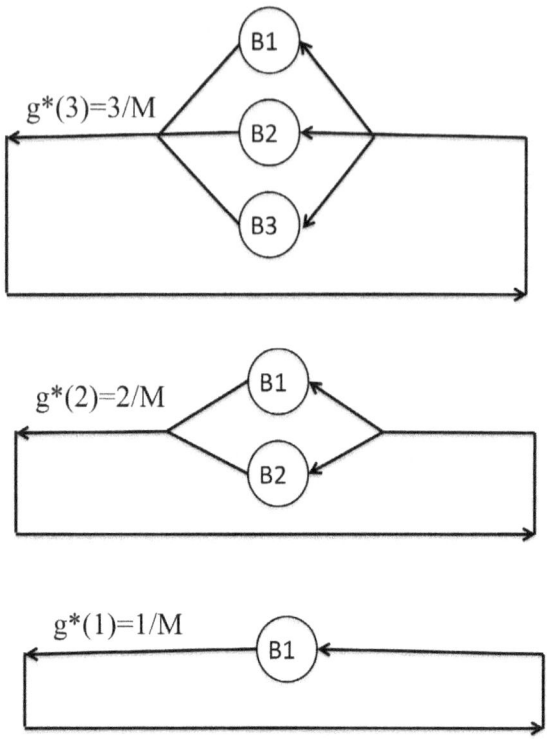

Figure 7-7. Off-line models for the original machines

To compute g*(2), note that the two tokens associated with servers B1 and B2 are entirely independent of each other as shown in the off-line model in the middle of Figure 7-7. There is no interference and no queuing delay in the off-line model. Thus g*(2) is equal to 2/M.

Exactly the same argument applies to g*(3). The three tokens in the off-line model at the top of Figure 7-7 circulate endlessly without encountering queuing delays, which implies that g*(3) is equal to 3/M.

Note that M is a directly observable quantity defined by equation (7-41). Thus the off-line values of g*(1), g*(2) and g*(3) are readily computable.

M = the mean time between failure during the observation interval

$$= \frac{3 \times T(0)^{\#} + 2 \times T(1)^{\#} + T(2)^{\#}}{A(0)^{\#} + A(1)^{\#} + A(2)^{\#}} \qquad (7\text{-}41)$$

7.10.2 Alternative model with variable rate server

The next step is to construct an alternative model by replacing servers B1, B2 and B3 by a variable rate server whose processing rate is a function of the number of customers present. The alternative model is shown on the left side of Figure 7-8. The variable rate server is identified by the arrow drawn diagonally across server B. The request completion rates for this server are set equal to g*(1), g*(2) and g*(3), which are the rates obtained from the off-line analysis.

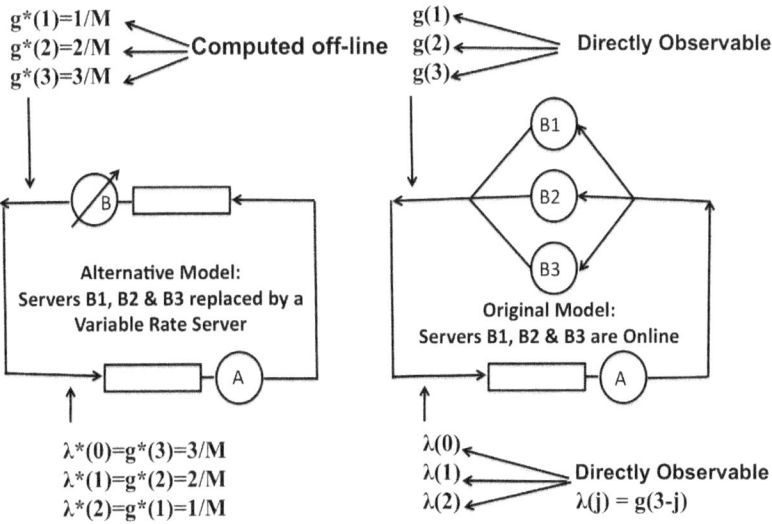

Figure 7-8. Replacing B1, B2 and B2 by a Variable Rate Server

The corresponding observable rates in the original model are g(1), g(2) and g(3). In this example, the assumption that online behavior equals off-line behavior implies that g(j)=g*(j) for j = 1, 2 and 3. This directly verifiable assumption may or may not be correct. If it is, the two models shown in Figure 7-8 are equivalent and the values of p(j) in the original model can be determined by solving the simpler alternative model, whose state transition diagram is shown in Figure 7-9.

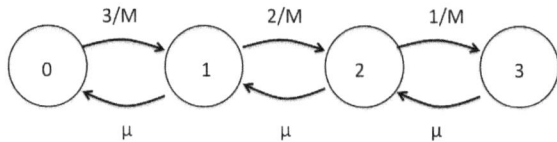

Figure 7-9. State transition diagram for the alternative model

The corresponding local balance equations are:

$$(3/M) \times p(0) = \mu \times p(1) \tag{7-42}$$

$$(2/M) \times p(1) = \mu \times p(2) \tag{7-43}$$

$$(1/M) \times p(2) = \mu \times p(3) \tag{7-44}$$

The solution is given by equations (7-45) through (7-48).

$$p(0) = \cfrac{1}{1 + \cfrac{3}{\mu M} + \cfrac{6}{(\mu M)^2} + \cfrac{6}{(\mu M)^3}} \tag{7-45}$$

$$p(1) = \frac{3}{\mu M} \times p(0) \tag{7-46}$$

$$p(2) = \frac{6}{(\mu M)^2} \times p(0) \qquad\qquad (7\text{-}47)$$

$$p(3) = \frac{6}{(\mu M)^3} \times p(0) \qquad\qquad (7\text{-}48)$$

Assuming the observable values of g(j) in the original model are equal to the computed values of g*(j) in the alternative model (i.e., assuming online behavior equals off-line behavior), it follows immediately that the solution given by equations (7-45) through (7-48) is also valid for the original model. This is the traditional solution to the machine repairman problem when the number of machines is equal to three. Standard texts on queuing theory present the general solution for an arbitrary number of machines.

7.10.3 Application to interactive computing

The machine repairman model was developed prior to the emergence of modern computer/communications systems. However, this same model is directly applicable to traditional mainframe-based timesharing and transaction processing systems, as well as modern web-based application servers and their back end database servers.

In all such cases, the repairman becomes the computer system being modeled, the machines on the factory floor become online users, and the time to repair an individual machine (i.e., $1/\mu$) becomes the time to process an individual transaction. The mean time between failure (i.e., M) is referred to as think time: it represents the amount of time the user spends examining the output of the last completed transaction and preparing the next. Submitting a transaction is the counterpart of a machine breaking down.

In the specific model illustrated in Figure 7-6, server A represents a computer system and servers B1, B2 and B3 represent online users. Each online user operates independently. The token that represents the status of an individual user cycles between the user's input device (desktop, laptop, tablet, etc.), the queue at server A, and the computer system represented by server A.

In more complex queuing models of computer/communications systems, the single repairman shown in Figure 7-6 is replaced by a complex sub-model that incorporates servers representing CPUs, I/O devices, channels, and so on. These extended models are discussed further in Section 7.11.2.

Regardless of the complexity of the sub-model incorporated within server A, the basic finite source nature of the arrival process remains the same: as more requests for service accumulate at server A, the number of active online users declines. As a result, the arrival rate at server A also declines. The main questions of interest in such cases concern the response time per transaction at server A and the overall throughput rate (i.e., completed transactions per second, per minute or per hour).

When the total number of users is large and server utilization is low to moderate, it is usually reasonable to approximate average response time for a finite source model of the type shown in Figure 7-6 by average response time for an infinite source model of the type shown in Figure 7-4. As already noted, the error introduced by this approximation is small in many cases of practical interest. (Buzen and Goldberg 1974)

One way to increase the productivity rate of the system illustrated in Figure 7-6 is to add a second repairman. This can be represented by the queuing model shown in Figure 7-10. Since there are now two repairmen, broken machines only have to wait in the queue when both repairmen are busy servicing other machines.

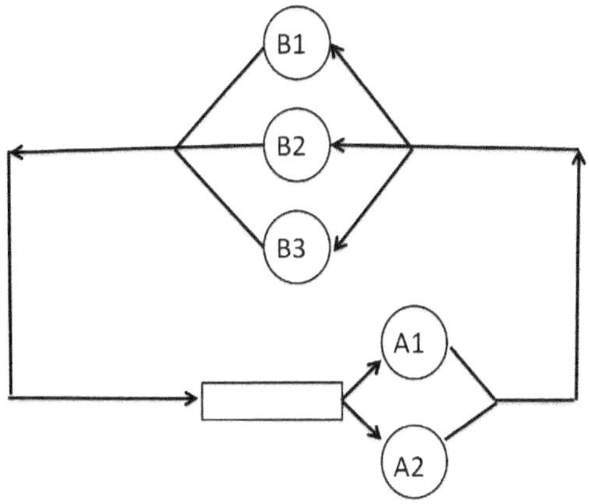

Figure 7-10. Model with two repairmen

The next step is to express the average repair time per breakdown as a function of observable quantities. The total number of breakdowns is the sum of $C(1)^{\#}$, $C(2)^{\#}$ and $C(3)^{\#}$. To compute the total time that is spent performing repairs, recall that $T(1)^{\#}$ is the total amount of time for which exactly one machine is broken. Only one repairman is working at such times. Similarly, $T(2)^{\#}$ is the total amount of time for which exactly two machines are broken, and $T(3)^{\#}$ is the total amount of time for which all three machines are broken. Both repairmen are working at these times. Thus, the total time spent performing repairs is:

$$T(1)^{\#} + 2 \times T(2)^{\#} + 2 \times T(3)^{\#}$$

The average amount of time per repair is

$$\frac{T(1)^{\#} + 2 \times T(2)^{\#} + 2 \times T(3)^{\#}}{C(1)^{\#} + C(2)^{\#} + C(3)^{\#}}$$

This implies that μ, the overall repair rate, is

$$\mu = \frac{C(1)^{\#} + C(2)^{\#} + C(3)^{\#}}{T(1)^{\#} + 2 \times T(2)^{\#} + 2 \times T(3)^{\#}} \tag{7-49}$$

Now consider the values of μ(j), which are the observed repair rates when the number of broken machines is equal to j. If the assumption that online behavior equals off-line behavior is satisfied, the values of μ(j) will have the following form.

$$\mu(1) = \mu \tag{7-50}$$

$$\mu(2) = 2\mu \tag{7-51}$$

$$\mu(3) = 2\mu \tag{7-52}$$

This leads to the state transition diagram shown in Figure 7-11.

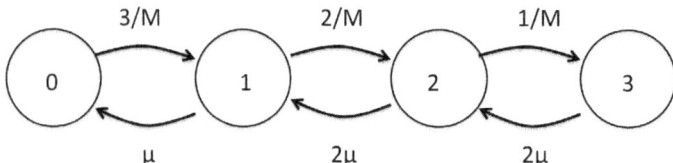

Figure 7-11. State transition diagram for two repairmen

The local balance equations for this state transition diagram are:

$$(3/M) \times p(0) = \mu \times p(1) \tag{7-53}$$

$$(2/M) \times p(1) = 2\mu \times p(2) \tag{7-54}$$

$$(1/M) \times p(2) = 2\mu \times p(3) \tag{7-55}$$

The solution to these equations is:

$$p(0) = \cfrac{1}{1 + \cfrac{3}{\mu M} + \cfrac{3}{(\mu M)^2} + \cfrac{3}{2 \times (\mu M)^3}} \qquad (7\text{-}56)$$

$$p(1) = \frac{3}{\mu M} \times p(0) \qquad (7\text{-}57)$$

$$p(2) = \frac{3}{(\mu M)^2} \times p(0) \qquad (7\text{-}58)$$

$$p(3) = \frac{3}{2 \times (\mu M)^3} \times p(0) \qquad (7\text{-}59)$$

It is interesting to compare this solution with the one obtained by replacing the original repairman by a new repairman who is twice as fast. Figure 7-12 presents the state transition diagram for this case. The values of p(j) can be obtained from equations (7-45) through (7-48): simply replace each occurrence of μ by 2μ.

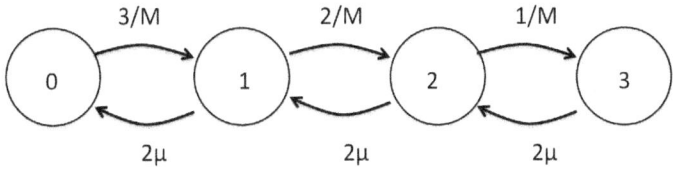

Figure 7-12. One repairman who is twice as fast

The application of this analysis to computer system performance should be self evident. Adding a second repairman corresponds to adding a second system or a second application server that is fed by the same queue of waiting transactions. Similarly, doubling the speed of the repairman corresponds to doubling the speed of the system represented by server A by upgrading its processors and other critical components. Queuing models are used routinely to analyze questions that arise in such settings.

7.11 Selected examples of practical significance

Queuing theory can be applied to many problems that arise during the design, analysis and day-to-day management of modern computer systems and communication networks. The examples in the last few sections have barely scratched the surface of this important subject. This section presents brief descriptions of three queuing models whose impact remains substantial more than four decades after their original development.

Solutions to the models discussed in this section are algebraically complex. This makes it difficult to derive useful intuitive insights by direct inspection of the solutions themselves. Nevertheless, it is important for practitioners to understand the assumptions that lie behind the corresponding models and the limitations that these assumptions impose.

7.11.1 Open queuing networks: ARPAnet and Internet

In certain situations, the output of one queue is linked to the input of another. Multiple queues connected in this manner form a queuing network. Figure 7-13 presents an example of such a network. Models of this particular type are referred to as open networks because customers are free to enter and leave: thus the number of customers actually in the network fluctuates over time.

As in the case of Figure 7-4, this open network is driven by infinite source arrival processes. In Figure 7-13, there are two infinite source queues on the left side of the diagram. Customers enter the network from these sources, flow through, and then exit on the right.

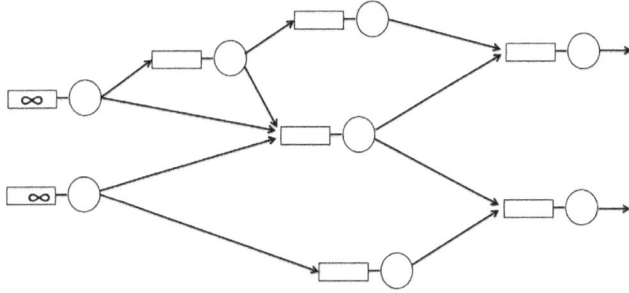

Figure 7-13. Model of a Store and Forward Network

Len Kleinrock (1962) pioneered the application of such models to the analysis of store and forward computer networks. His work has had a major impact on the modern world. Today's Internet is a direct descendant of the ARPAnet, a computer network developed in the late 1960s under sponsorship of the Department of Defense Advanced Research Projects Agency (ARPA). During the earliest phases of this project, Kleinrock's models were used to determine whether or not such a network would ever be able to deliver satisfactory levels of performance. The decision to move ahead with funding for the project depended on the results of Kleinrock's analysis, which was carried out before a single line of code was written (Severance 2014). Later, Kleinrock's lab at UCLA hosted the first node on the ARPAnet and was the point of origin for the first packets actually transmitted over this network.

The analysis in Kleinrock's PhD dissertation includes a highly detailed queuing model that keeps track of the properties of individual messages as they move from node to node. Representing message flow at this level of detail leads to a model of enormous mathematical complexity. To obtain a closed form analytic solution, Kleinrock introduced a profoundly important simplifying assumption. In essence, this assumption enabled each node to be separated from the rest of the network and modeled as an independent queuing server.

As part of his PhD dissertation, Kleinrock also constructed a detailed simulation model that actually tracked the identities of individual messages. When he compared the results of his detailed simulation with the approximate analytic model, he discovered that the differences were relatively small and that the additional details incorporated into the simulation model did not have a material effect on the main results of his analysis. From the perspective of observational stochastics, Kleinrock's simplifying assumption has much in common with the assumption that online behavior is equal to off-line behavior: both lead to acceptably accurate queuing models even though certain details are not represented precisely.

7.11.2 Closed queuing networks: computer performance

Figures 7-2 and 7-6 are examples of closed queuing networks. A fixed number of tokens (customers) circulate around these networks at all times. There are no infinite source servers capable of injecting new customers into the network; conversely, customers who are in the network at time zero have no way to exit.

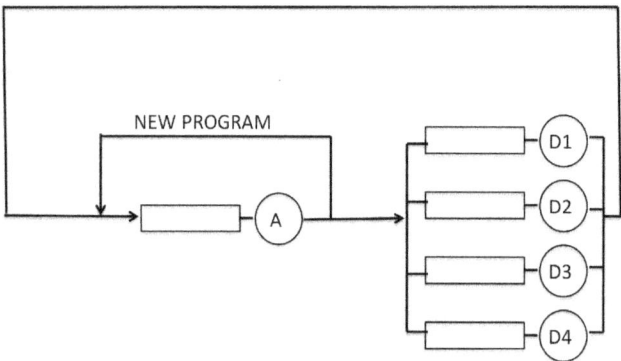

Figure 7-14. Central server model

Figure 7-14 illustrates an especially useful closed queuing network known as the central server model (Buzen 1971, 1973). This model represents a multiprogrammed computer system operating under a steady backlog. Each token circulating around this network represents an active program. Program behavior is characterized as an alternating sequence of CPU processing bursts and I/O transfers. When a program terminates, its associated token moves along the "new program" loop and returns representing the first processing burst of the next program waiting in the backlog. Thus the rate at which tokens flow through the new program loop represents the throughput of the system being modeled.

Only four I/O devices are illustrated in Figure 7-14, but the model can be extended to include any desired number of devices. There is also just one CPU in Figure 7-14, but multiple CPUs (or multi-core CPUs) can be accommodated by employing the technique used in Figure 7-10 to represent multiple repairmen. The number of active programs circulating through the model is linked directly to the amount of available main memory. This makes the central server model well suited for analyzing the benefits of increasing memory size, upgrading processor speed or the number of tightly coupled processors (SMPs), and upgrading I/O device performance by installing faster devices, cached storage controllers, or solid state disks. The central server model is particularly useful for identifying bottlenecks and for assessing alternative strategies for alleviating these bottlenecks.

The central server model can also be modified to represent computer systems that process workloads driven by online users working at their terminals. Simply replace the new program loop by a set of interactive terminals as shown in Figure 7-15. The interactive terminals behave in the same manner as the finite source arrival process illustrated in Figure 7-6.

A generalized closed form solution for queuing models of the type illustrated in Figures 7-14 and 7-15 was originally derived by Jackson (1963) and, independently, by Gordon and Newell (1967).

The solution constitutes a major milestone in queuing theory. However, for a closed model with N circulating customers and M servers, the number of possible states is equal to $\dfrac{(M+N-1)!}{N!(M-1)!}$.

This number grows explosively as M and N increase, making direct evaluation of the closed form solutions derived by Jackson and by Gordon and Newell computationally intractable for all but the simplest of models. As a result, these solutions attracted little attention from researchers and practitioners for several years.

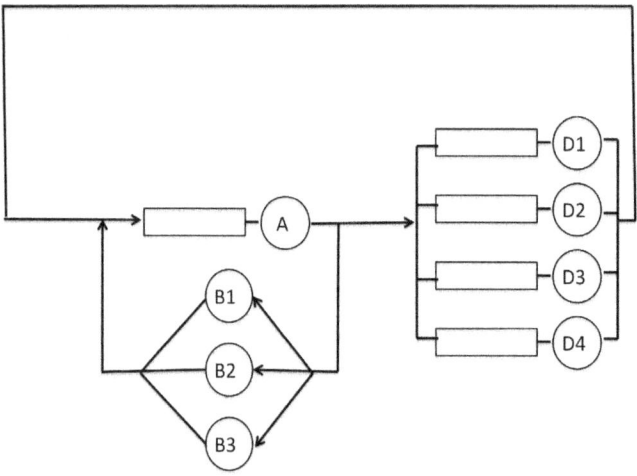

Figure 7-15. Central server model with interactive users

The situation changed abruptly in the summer of 1970 with the discovery of a powerful set of recursive relationships among the elements of the closed form solutions. These relationships provided the basis for remarkably efficient iterative algorithms for evaluating Gordon and Newell's general solution (Buzen 1971, 1973). These same recursions also yielded simple algebraic expressions for related quantities that are difficult to evaluate directly: for example, the marginal queue length distribution for each server in a queuing network (which can then be used to compute mean queue lengths, percentiles, etc.). See Bruell and

Balbo (1980, pp. 36-55), Harrison and Patel (1993, pp. 235-238), Bolch, Greiner, de Meer and Trivedi (2006, pp. 369-384), Stewart (2009, pp. 570-581), or Gelenbe and Mitrani (2010, pp 106-114) for extended discussions of this approach, now known as the convolution algorithm or Buzen's algorithm (Wikipedia).

In combination with Jackson's general solution, the central server model and the availability of efficient computational algorithms ignited a surge of activity by an international community of researchers. This led to a rapid succession of major advances in the theory of queuing networks and to the development of other efficient computational algorithms for the evaluation of these networks.

These advances, which are described in the texts cited in the preceding paragraph and in other texts cited in the Preface, were soon incorporated into commercially successful modeling tools (Buzen, Goldberg, Langer, Lentz, Schwenk, Sheetz and Shum 1978). Tools of this type have been used for decades to manage performance and plan capacity at many of the world's largest data centers (Casale, Gribaudo and Serazzi 2010). Queuing network models remain important tools in the modern era of cloud computing, service oriented architecture, resource virtualization and big data.

Note: The "product form" solutions for closed queuing networks that were obtained by Jackson (1963) and by Gordon and Newell (1967) were derived under traditional stochastic assumptions. Solutions having the same algebraic form can also be derived within the framework of observational stochastics. The most important assumption required for this derivation is that the online behavior of each server in the network must be equal to its off-line behavior. Certain empirical independence assumptions are also required. See Denning and Buzen (1978) for further details.

7.11.3 The ALOHA network: origins of Ethernet

The ALOHA system was an early computer network designed by Norm Abramson of the University of Hawaii (Abramson 1970). In this network, user terminals on separate islands communicated with a centrally located computer system via a radio link. All terminals shared a single UHF channel for sending inbound packets. A second channel was used at the central site for outbound responses.

One of the most challenging problems in this environment is managing access to the shared channel used for inbound packets. Since the terminals are geographically remote, there is no easy way to coordinate this access. Abramson's solution is noteworthy for its simplicity and its audacity: the basic idea is to allow each terminal to operate with complete disregard for what other terminals may or may not be doing. In this chaotic environment, some packets get through; others are garbled because they are sent at times when the channel is already busy. To accommodate such failures, the central system sends back an explicit acknowledgement when an undamaged packet is received. If a terminal does not receive an acknowledgement within a specified timeout window, the terminal assumes that a collision has occurred and retransmits the packet.

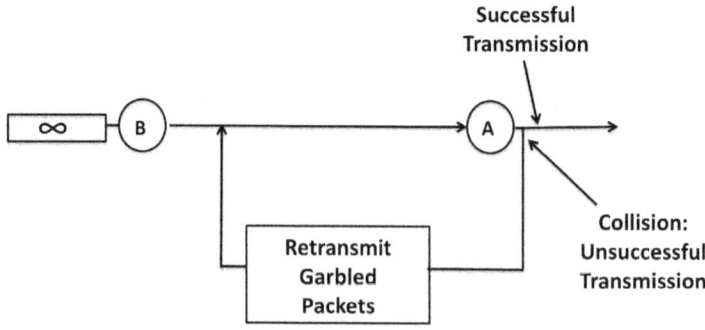

Figure 7-16. Original ALOHA model

Abramson developed the open queuing model sketched in Figure 7-16 to explore the behavior of this network. Server A represents the UHF channel used for inbound packets. Service time, which is the average time required to transmit one packet through this channel, is a function of channel bandwidth and packet length. If a second packet arrives when one is already being transmitted, a collision occurs and both packets are garbled. They must then be retransmitted, perhaps multiple times, before they finally get through. The retransmission process is represented by the box at the bottom of the diagram.

Under these general assumptions, Abramson constructed a queuing model that he used to determine the maximum effective utilization of a basic ALOHA channel. [Effective utilization is utilization due to newly arriving packets. It does not include the load generated by retransmitted packets.] Abramson showed that the maximum effective utilization of a basic ALOHA channel is equal to $1/2e$ (approximately 18%), where e is the mathematical constant 2.718... .

Only one aspect of Abramson's model is important for this discussion: his model incorporates an infinite source arrival process of the type depicted in Figure 7-3. Each terminal in Abramson's model generates an independent stream of new packets. All terminals generate packets at the same basic rate. Thus, if there are N terminals, the total arrival rate for new packets is N times this basic rate. An assumption of this type is not uncommon and can be found in many queuing models.

Bob Metcalfe recognized that this assumption has certain undesirable implications when applied to the analysis of "what if" questions that involve increasing the number of terminals. Imagine a case where terminals are added (one by one) to an ALOHA network. As long as effective utilization remains below $1/2e$, the network will be stable and Abramson's model will yield a well defined result. However, at some point adding the next additional terminal will cause the total effective utilization to exceed 1/2e. At this point, the model predicts that the channel will become unstable due to an "uncontrolled regenerative burst of retransmission." (Metcalfe 1973, Chapter 5)

Based on his practical experience with networks, Metcalfe regarded such a catastrophic collapse (triggered by the addition of a single terminal) as unlikely and unrealistic. He believed that ALOHA networks will remain stable as each new terminal is added and that the number of terminals in retransmission mode – and thus the total number of retransmissions per second - will increase gradually (rather than abruptly) during this process.

Metcalfe went on to develop a queuing model that predicts this type of behavior for his PhD dissertation. (Metcalfe 1973) In his revised ALOHA model, the load on the network is generated by a finite source arrival process of the type depicted in Figure 7-6. Abramson's limit of 1/2e for effective utilization is still applicable, but this limit is now approached in a smooth asymptotic fashion as the number of attached terminals increases. The overall structure of Metcalfe's ALOHA model is sketched in Figure 7-17.

During the course of his analysis, Metcalfe recognized that the performance of ALOHA-like networks could be improved substantially by testing the status of the channel immediately prior to sending a packet, and by only transmitting packets when the channel is free. This testing procedure is known as carrier sensing. It reduces, but does not completely eliminate, the proportion of packets that collide with one another and have to be retransmitted. Collisions can still occur if two terminals detect a "not busy" condition at the same time and then decide to initiate transmissions concurrently.

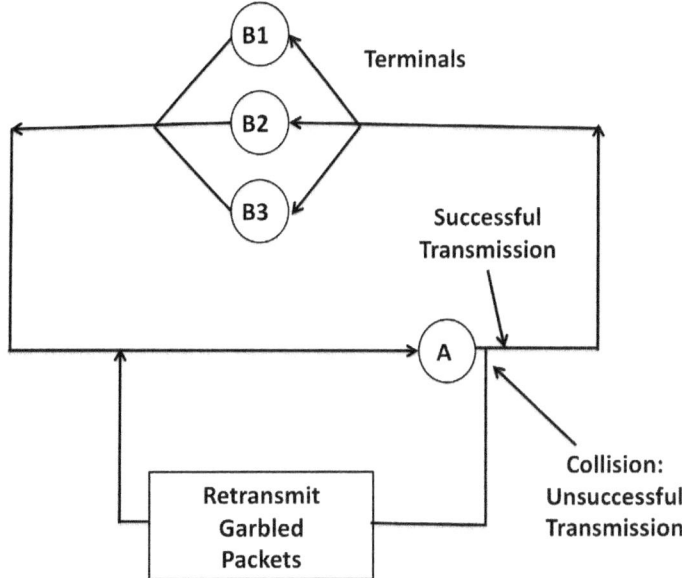

Figure 7-17. Metcalfe's ALOHA model

Metcalfe's research on ALOHA led to another important insight: he recognized that ALOHA-like networks augmented with carrier sensing would be especially useful in settings where terminals are relatively close geographically and are connected by physical wires rather than a radio channel. The Ethernet protocol, which is the *de facto* standard for most of today's local area networks, began as a practical application of this insight.

Bob Metcalfe successfully defended his PhD dissertation with its refined model of the ALOHA network in May 1973. The following week he returned to California and coined the term Ethernet in an internal memo written at Xerox PARC (Palo Alto Research Center). Metcalfe and his colleague David Boggs had the first Ethernet LAN up and running at PARC six months later.

There is, of course, much more to this story (Metcalfe and Shustek 2006), but it remains a compelling example of the link between queuing models and inventions that have had a profound impact on the modern world.

CHAPTER 8

Proportion, Probability and Chance

8.1 The proportionalist view of probability

The problems considered in observational stochastics are phrased most naturally in terms of proportions rather than probabilities. For example, practitioners are typically concerned with the proportion of time a server is busy rather than the probability a server is busy. Similarly, practitioners are typically concerned with the proportion of packets that arrive when a buffer is already full rather than the probability of a buffer being full when a packet arrives.

This chapter examines the use of proportions in contexts where the passage of time is not an integral element of the original problem, and where problems are phrased most naturally in terms of questions that begin with "what is the chance" or "what is the likelihood." Note that the validation paradigm described in Section 1.2 is not directly applicable to such contexts since there are no time-dependent trajectories to observe. Instead, there are only static populations. In this context, intuitive concepts such as chance or likelihood become meaningful and measurable when linked to directly observable proportions of these populations. As in the case of observational stochastics, the structure of these populations is typically characterized through loose constraints rather than probability distributions and the sampling premise.

The next few sections illustrate how this proportionalist view can be employed to formalize and analyze questions that are expressed using ordinary non-technical terms such as chance or likelihood. As usual, a series of simple examples will be used to illustrate fundamental principles and concepts that extend broadly.

8.2 A medical test and its implications

This example, which is adapted from Example 2.10 in (Kobayashi, Mark and Turin 2012), concerns a medical test that is used to diagnose a particular disease. The original statement of the problem begins as follows:

"The following information is known about this disease and its medical test.

1. For a person with this disease, the test yields a positive result 99% of the time and a negative result 1%.

2. For a person without this disease, the test yields a negative result 99% of the time and a positive result 1%.

3. 1% of the population is infected by this disease and 99% of the population is not."

Note that Assumptions 1, 2 and 3 are expressed in terms of relationships among observable proportions of a population. Their meaning can be understood, quantified and verified without referring to abstract entities such as sample spaces, events, probability measures and random variables.

The example continues by assuming that an individual has taken the test and obtained a positive result. In other words, the test indicates that this individual has the disease. However, it is possible that the individual is perfectly fine, but the test has produced a false positive. Alternatively, it is also possible that the test is accurate and the individual is indeed ill.

Within this context, Kobayashi Mark and Turin propose the following question: "What is the chance" that this particular individual has the disease?

8.2.1 Alternative formulations and solutions

This problem appears immediately following a derivation of Bayes' Theorem. Kobayashi, Mark and Turin proceed by applying the theorem to obtain a solution. As part of this process, the authors implicitly restate the original problem and the associated assumptions in terms of mathematical probabilities. The restated problem is: what is the probability that this particular individual has the disease? The restated assumptions are presented below as A, B and C. None of these restatements actually appear in (Kobayashi et al. 2012), but they are implicit in the analysis.

A.   The probability of a positive test result, given that the individual has the disease, is equal to .99

B.   The probability of a positive test result, given that the individual does not have the disease, is equal to .01

C.   The unconditional probability that the individual has the disease is equal to .01

Assumptions A, B and C transform the original problem regarding aspects of real world entities (populations, test results and rates of disease) into a problem within the realm of probability theory. As shown by Kobayashi et al., this problem can be solved by an application of Bayes' Theorem.   This leads directly to the following conclusion: the probability that an individual who tests positive actually has the disease is equal to one half. Section 8.5 provides the step-by-step details of this traditional probabilistic analysis.

This section presents an alternative analysis based on the proportionalist view of probability. The first step is to formalize Assumptions 1, 2 and 3 using the previously introduced notational conventions of observational stochastics. In this case, the hash tag is again used to identify formal variables that correspond to directly observable raw counts. These raw counts are associated with static populations rather than dynamic trajectories. In effect, populations can be regarded as trajectories whose elements are not ordered with respect to time. In all other respects, populations and trajectories are identical.

$N^{\#}$ = number of individuals in the entire population

$D^{\#}$ = number of individuals who have the disease

$W^{\#}$ = number of individuals who are well (i.e., do not have the disease)

$$= N^{\#} - D^{\#} \qquad\qquad (8\text{-}1)$$

$Dpos^{\#}$ = number of individuals who have the disease and test positive

$Wpos^{\#}$ = number of individuals who are well (i.e., do not have the disease) and test positive

Assumptions 1, 2 and 3 can now be stated formally as follows:

A = Proportion of accurate test results for those individuals who have the disease

$$= Dpos^{\#} / D^{\#}$$

$$= .99 \qquad\qquad (8\text{-}2)$$

I = Proportion of inaccurate test results for those individuals who are well

$$= Wpos^{\#} / W^{\#}$$

$$= .01 \qquad\qquad (8\text{-}3)$$

D = Proportion of individuals who have the disease

$$= D^{\#} / N^{\#}$$

$$= .01 \tag{8-4}$$

Now consider the original question: what is the chance that an individual who tests positive actually has the disease? While this question is meaningful on an intuitive level, it must be re-expressed in terms that are both observable and quantifiable before the analysis can continue.

To motivate this next step, begin with a simpler case where the only information available is that 1% of the population is infected by this disease, implying that 99% of the population is not. In this simple case, it is reasonable to assume that the chance the individual has the disease is 1% or .01. In other words, if an individual is informed that he or she is a member of this population, and if this is the only information available, it is reasonable for this individual to assume that his or her chance of having the disease is .01.

More generally, the chance that a member of a population has a certain attribute will be defined as the proportion of the population that has that attribute. This definition, which forms the foundation for the proportionalist perspective, can neither be proven nor disproven through a formal analysis. However, it is intuitively satisfying and, as noted in Chapter 1, compatible with Poincaré's classic definition (Poincaré 1905): "The probability of an event is the ratio of the number of cases favorable to the event to the total number of possible cases."

Returning to the original question, what is the chance that an individual who tests positive actually has the disease? The answer is obtained by considering all individuals who have tested positive and then determining what proportion of this population actually has the disease. This proportion represents the chance that such an individual is ill.

Note that the total number of individuals who test positive is equal to the sum of $\text{Dpos}^\#$ and $\text{Wpos}^\#$. Of these individuals, the number who are actually ill is equal to $\text{Dpos}^\#$. Thus, from the proportionalist perspective, the chance that an individual who tests positive is actually ill can be represented by the variable X in equation (8-5)

$$X = \frac{\text{Dpos}^\#}{\text{Dpos}^\# + \text{Wpos}^\#} \qquad (8\text{-}5)$$

To solve the original problem it is necessary to express X as a function of A, I and D.

By equation (8-2),

$$\text{Dpos}^\# = A \times D^\# \qquad (8\text{-}6)$$

By equations (8-1) and (8-3),

$$\text{Wpos}^\# = I \times W^\#$$

$$= I \times (N^\# - D^\#) \qquad (8\text{-}7)$$

Replacing $\text{Dpos}^\#$ and $\text{Wpos}^\#$ in equation (8-5) by the values in equations (8-6) and (8-7) yields equation (8-8).

$$X = \frac{A \times D^\#}{A \times D^\# + I \times (N^\# - D^\#)} \qquad (8\text{-}8)$$

Dividing numerator and denominator of the right hand side of equation (8-8) by $N^{\#}$ and using the fact that $D^{\#} / N^{\#} = D$,

$$X = \frac{A \times D}{A \times D + I \times (1 - D)} \qquad (8\text{-}9)$$

This is the general solution. When $A = .99$ and $D = I = .01$,

$$X = \frac{.99 \times .01}{.99 \times .01 + .01 \times .99} = .50 \qquad (8\text{-}10)$$

This is the same result derived in Section 8.5 using Bayes' Theorem.

8.2.2 Frequentist and Bayesian perspectives

Kobayashi, et al. follow their analysis with a discussion of how the mathematical solution to this problem can be interpreted in practice. Essentially, the discussion concerns the meaning of the term probability. In the preceding example, the probability that an individual with a positive test result actually has the disease is equal to one half. Frequentists and Bayesians have different views regarding the practical implications of this finding.

The frequentist view of probability is based on the idea of conducting a large number of repeated trials and observing the proportion of those trials that produce a certain result. In this case, the result is that the individual who tests positive does in fact have the disease. This notion of independent trials is well suited for the analysis of games of chance, but is awkward to apply in this case since there is only one individual and one test result.

Kobayashi, Mark and Turin suggest that frequentists might interpret the notion of independent trials as being equivalent to examining a large set of "individuals with background similar to the individual in question." Each examination can, in a sense, be construed as an independent trial. Assuming that such a set of individuals can be identified and diagnosed correctly, a frequentist would expect to find the disease afflicting half the individuals in the set – provided the size of the set is very large.

The authors then consider the same solution from a Bayesian perspective. In this framework, probability is a subjective concept related to the degree of belief that a certain outcome is true. There is no need to rely upon the notion of a repetitive sequence of independent trials, and no need to assume that uncertainty is the result of an intrinsically random process.

Since one percent of the population has the disease according to Assumption 3, the authors note that a Bayesian is likely to assign a value of .01 to the subjective probability (i.e., belief) that an individual has the disease. This value, which is assigned before any information regarding the individual's test result becomes available, is referred to as a prior probability. Once additional information (i.e., evidence) regarding the individual's test result becomes known, Bayesians use this evidence to calculate the posterior probability that the individual has the disease. This leads to a solution with the same form as equation (8-9). Under Assumptions 1, 2 and 3, the posterior probability that the individual has the disease is then equal to one half.

It is interesting to note that Kobayashi, Mark and Turin employ the notion of observable proportions in their discussions of both perspectives. For frequentists, the probability of an event is linked to the observable proportion of occurrences of the event during a large number of repeated trials. For Bayesians, the subjective probability of an event is linked – at least in this case - to the observable proportion of occurrences of the event in the entire population (Assumption 3) or in the sub-population comprised of individuals with positive test results (equation (8-9)). Subjective probability is not always linked to observable proportions, but in this example it is reasonable to assume that it is.

8.2.3 Poincaré and the proportionalist perspective

The proportionalist approach being developed in this chapter also relies strongly on observable proportions. Following Poincaré (1905) and many others who have considered this issue, the chance or probability associated with an event is defined simply as the proportion of occurrences of that event within a population.

The medical test has provided a concrete example of this approach. Several different proportions are involved. In equations (8-2), (8-3) and (8-4), the directly observable variables A, I and D are all defined as proportions. The chance that an individual who tests positive actually has the disease, which is represented by X, is also defined as a directly observable proportion through equation (8-5).

Note that the observable value of X is always predicted with absolute certainty by equation (8-9), provided the values of A, D and I are known. This is analogous to the level of certainty that characterizes results derived through observational stochastics. Remember that the goal of this analysis is not to predict whether or not an individual actually has the disease. Rather, the goal is to determine the chance that the individual has the disease. Chance is defined as the proportion of individuals who have the disease within a population or sub-population of individuals who are similar to the individual in question.

For the practitioner, the main benefit of dealing with simple proportions rather than generalized probabilities is the direct relationship between derived results and observable data. Focusing the analysis on observable proportions also legitimizes straightforward arguments of the type used to derive equation (8-9). These proportionalist arguments do, however, presume that all observable values are finite. This limitation is of no concern to most practitioners since infinite values never arise when measuring properties of entities that exist in the real world. See Section 8.6.

The full statement of Poincaré's characterization of probability in Chapter 11 of *Science and Hypothesis* (1905) includes an additional consideration: "The probability of an event is the ratio of the number of cases favorable to the event to the total number of possible cases, ... provided the cases are equally probable."   To understand why Poincaré adds this provision, note that ratios and proportions are always computed over populations or sub-populations. If some of the possible cases within these populations or sub-populations are more probable than others, the intuitive notion of chance must be refined.   It cannot be equated with a simple ratio or proportion.

Poincaré illustrates this point with a simple example: suppose a pair of dice are tossed across a table: what is the chance that at least one of the dice will turn up a six?   The standard way to analyze this problem is to treat each die as a distinguishable physical object that can turn up in any of six possible ways. Thus, when two dice are tossed, the total number of possible outcomes is $6 \times 6 = 36$. Of these 36 possible outcomes, 11 contain at least one *six*. This implies that the chance of getting at least one six is 11/36.

The alternative analysis begins by representing the outcome of each toss as a set {a,b} where $a \leq b$, and where both a and b can range from 1 to 6.   This characterization, which is entirely legitimate, treats the two dice as indistinguishable.   The total number of possible distinct outcomes is no longer 36. Instead it is only $6 \times 7 / 2 = 21$. Under this form of representation, the number of favorable outcomes that contain at least one *six* is equal to six:

$$\{1,6\}, \{2,6\}, \{3,6\}, \{4,6\}, \{5,6\}, \{6,6\}$$

There is nothing wrong with the analysis up to this point. However, Poincaré points out that it is incorrect to use this information to conclude that the chance of a favorable outcome is 6/21. The problem with this conclusion is that all 21 possible outcomes are not equally probable. For example, the outcome {5,6} is twice as probable as the outcome {6,6} since there are two ways to generate the former and only one way to generate the latter. Poincaré's requirement that all possible cases must be equally probable implies that the second analysis is not correct.

For the medical test, the population used in the computation of X is the set of all individuals who have taken the medical test and have received a positive result. Since nothing else is known about this population, it is entirely reasonable to believe that each member has the same likelihood of being ill. This satisfies Poincaré's requirement that all possible cases must be equally probable.

Note that Poincaré's notion of cases being equally probable is a subjective idea that reflects the underlying beliefs of the analyst. In this sense it is similar to the Bayesian notion of subjective probability: in particular the assumption of a uniform prior distribution. It is also linked to the frequentist concept of conducting a large number of equivalent experiments and then computing the proportion of time a specific outcome is observed. This procedure can only be justified if the specific outcome has the same chance of being observed in each experiment: in other words, if the outcome is equally probable in all experiments.

The assumption of equally probable outcomes is also closely linked to the intuitive notion of selecting a sample at random from a finite population. In such situations, the assumption of random selection is equivalent to the assumption that all possible cases (i.e., all possible samples) are equally probable. The next few sections present a series of examples that illustrate this perspective.

8.3 Dealing cards and drawing samples

This particular series of examples begins with a trivial question: what is the chance that a card drawn at random from a conventional deck of cards is an ace? In this example the population consists of the 52 cards in the deck. This population exhibits all the standard properties one would expect, including the fact that the number of aces in the deck is equal to four.

Since four of the 52 cards in the deck have the property of being an ace, the chance of drawing an ace is simply $4/52 = 1/13$. This follows directly from the proportionalist definition of chance presented in Section 8.2.1. There is no need to represent each individual card in the deck by a random variable or to associate a probability distribution with the value of a particular card that has been drawn but has not yet been inspected. From a proportionalist perspective, the fact that there are four aces in a deck of 52 cards is all that really matters.

Now consider the chance of being dealt a royal flush (i.e., 10-J-Q-K-A of the same suit) during a game of poker. In this situation, the population is not the deck itself. It is instead the set of all possible five-card hands that can be dealt from the deck. Each hand can be regarded as a sample that has been drawn "at random" from the deck. Thus, the set of all possible hands can be regarded as constituting an equally probable pool of samples. This sample pool is the population for the analysis.

If the deck is well shuffled and the dealer is honest, it is reasonable to assume that each member of this pool has the same chance of being dealt. This implicit assumption conforms once again to the proportionalist interpretation of chance.

The number of distinct hands in the pool can be calculated by standard combinatoric arguments: simply divide the product

$52 \times 51 \times 50 \times 49 \times 48$ by $1 \times 2 \times 3 \times 4 \times 5$. This yields a total of 2,598,960 distinct hands. It is clear that only four of these hands are royal flushes. Thus, from a proportionalist perspective, the chance of being dealt a royal flush is 4 divided by 2,598,960 which is equal to 1/649,740.

Many other problems involving chance can be formalized and solved using this straightforward proportionalist approach. All these cases involve samples that are drawn at random from a population. This makes it reasonable to assume that all samples are equally probable. In each case, the mathematical challenge is to determine two quantities: the total number of equally probable samples in the sample pool; the number of samples in that pool that have a certain attribute. Dividing the former by the latter yields the chance that a sample has the desired attribute.

8.4 Predicting the outcome of an election

The problems described in the preceding section are all relatively simple. A more interesting and challenging class of problems arises when the relationship between populations and samples is inverted. Rather than considering the chance that a hand dealt from a deck of cards has a certain attribute (e.g., being a royal flush), consider instead the chance that the entire deck has a certain attribute, given that a specific hand has actually been dealt. In general terms, the challenge is to make inferences about an entire population, given information that has been observed in one or more samples.

8.4.1 Samples and sample pools

The sample pools that were introduced in the previous section play an important role in the analysis of such problems. However, additional refinements are required to deal with these cases. These refinements can – once again – be illustrated with the aid of a simple example.

Suppose an analyst wishes to determine the chance that candidate A will win an election. To simplify the analysis, assume there are only nine voters in the population. Suppose a sample of three voters is selected at random from this population. All three voters favor candidate A (rather than candidate X). What is the chance candidate A will win?

Assuming there are no undecided voters, the number of voters who favor candidate A is either 0, 1, 2, 3, 4, 5, 6, 7, 8 or 9. One way to proceed is to consider which of these alternatives is most likely, given that all three voters in the sample favor candidate A. If the number of voters in the population who favor candidate A is equal to nine, all samples of size three are certain to contain three A votes. On the other hand, if there are fewer than nine A voters in the population, it is always possible for a sample to contain at least one vote for candidate X. Thus, given that the sample contains three A votes, the most likely alternative is that all nine voters are in favor of candidate A.

The analysis in the proceeding paragraph is correct but clearly fails to account for the subtle nature of the original question. The problem is not to determine the most likely number of voters who favor candidate A. Instead, the problem is to determine the chance that candidate A will win. This will happen whenever five or more voters favor candidate A. Each of these cases must be considered separately to carry out a proper analysis.

Continuing with this reasoning, the next step is to determine the chance that the sample contains three A votes when n, the total number of A voters in the entire population, is equal to 0, 1, 2, 3, 4, 5, 6, 7, 8 or 9. These ten possible values of n form the top row of Table 8-1.

Each of these cases can be evaluated from a proportionalist perspective. Note first that the number of distinct samples of size three that can be drawn from a population of nine voters is equal to $(9 \times 8 \times 7)/(1 \times 2 \times 3) = 84$. These 84 distinct samples constitute the sample pool. As shown on the second row of Table 8-1, this sample pool is the same for each value of n.

0	1	2	3	4	5	6	7	8	9
84	84	84	84	84	84	84	84	84	84
0	0	0	1	4	10	20	35	56	84
$\dfrac{0}{84}$	$\dfrac{0}{84}$	$\dfrac{0}{84}$	$\dfrac{1}{84}$	$\dfrac{4}{84}$	$\dfrac{10}{84}$	$\dfrac{20}{84}$	$\dfrac{35}{84}$	$\dfrac{56}{84}$	$\dfrac{84}{84}$

Table 8-1. Chance all three voters in the sample favor candidate A

The number of samples within each pool that contain exactly three votes for candidate A can once again be calculated by standard combinatoric arguments. For each value of n, this number is:

$$\frac{n \times (n-1) \times (n-2)}{1 \times 2 \times 3} = n \times (n-1) \times (n-2)/6 \qquad (8\text{-}11)$$

These values appear in the third row of Table 8-1. Assuming that all 84 members within a given sample pool are equally probable, the chance that a sample has a certain attribute (for a specific value of n) is defined as the proportion of samples in the associated sample pool with that attribute (in this case, the proportion of samples that contain exactly three voters who favor candidate A). Thus the chance the sample contains exactly three voters in favor of candidate A for each possible value of n is shown in the last row of Table 8-1.

To complete the analysis, it is necessary to use the information in Table 8-1 to determine the overall chance that candidate A will win: in other words, to determine the chance that the total number of A voters in the entire population is 5, 6, 7, 8 or 9.

All the samples enumerated in the third row of Table 8-1 contain exactly three voters who favor candidate A. The total number of such samples is: $1 + 4 + 10 + 20 + 35 + 56 + 84 = 210$

Since the original sample contains exactly three votes for candidate A, that sample must appear as one of the 210 samples in row three. Of these samples, 205 correspond to cases where $n > 4$. If all 210 samples are equally probable, it can thus be argued that the chance candidate A will win is 205/210.

8.4.2 Equally probable sub-pools

The assumption that all 210 samples in row 3 have the same chance of being selected is reasonable under some circumstances, but not others. To examine this issue further, note that each column in Table 8-1 summarizes the properties of 84 possible samples that can be drawn when the number of votes for candidate A has a fixed value. Since there are ten columns, the total number of distinct samples summarized in this table is $84 \times 10 = 840$. These 840 possible samples form the sample pool for the discussion in this section.

It is helpful to assign a unique identifier to each of these 840 distinct samples. First assign the labels A, B, C, D, E, F, G, H and I to the nine voters in the population. Each possible sample of three voters can then be represented by a unique string of three letters in alphabetical order: ABC, ABD, ABE, etc. As already noted, there are 84 such strings.

Next append a tag (i.e., a single digit) to the end of each string. The value of this tag (i.e., 0, 1, 2 ... 9) represents the number of voters who favor candidate A in the sub-pool from which the sample has been drawn. A set of samples with the same tag will be called a sub-pool. Thus, DEF4 identifies the sample generated when voters D, E and F are selected from the sub-pool in which the total number of voters who favor candidate A is equal to 4. Similarly, DEF9 identifies the sample generated when the same three voters are selected from the sub-pool in which the total number of voters who favor candidate A is equal to 9.

Samples DEF4 and DEF9 may not have the same chance of being selected, even though both samples contain the same three voters. However, they will be equally probable if the overall chance of drawing a sample from sub-pool 4 is equal to the overall chance of drawing a sample from sub-pool 9. In general, all 840 random samples in the entire pool will have the same chance of being selected if all ten sub-pools are equally probable. Under these assumptions, Poincaré's concept of all outcomes being equally probable is appropriate, and the chance candidate A will win is equal to 205/210.

8.4.3 Replication of members in sub-pools

To understand why DEF4 and DEF9 might not be equally probable, suppose a review of the past elections reveals that the vast majority of elections are decided by a single vote. In particular, suppose that elections where candidate A looses by a margin of 4 to 5 or wins by a margin of 5 to 4 are ten times more common than any of the other possible margins.

Projecting this finding into the future, suppose it is reasonable to assume that the chance of drawing DEF4 from sub-pool 4 is ten times greater than the chance of drawing DEF9 from sub-pool 9. Similarly, suppose it is also reasonable to assume that the chance of drawing DEF5 from sub-pool 5 is ten times greater than the chance of drawing DEF9 from sub-pool 9.

Assumptions of this type can be represented within the proportionalist framework by adding duplicate members to sub-pools whose members have greater chances of being selected. In this case, the sample DEF4 can be replaced by ten duplicate samples: DEF40, DEF41, DEF42 … DEF49. The same rationale can be applied to the other 83 samples that make up the sub-pools associated with columns 4 and 5 of Table 8-1. The result of these tenfold duplications is shown in Table 8-2.

0	1	2	3	4	5	6	7	8	9
84	84	84	84	840	840	84	84	84	84
0	0	0	1	40	100	20	35	56	84
$\frac{0}{84}$	$\frac{0}{84}$	$\frac{0}{84}$	$\frac{1}{84}$	$\frac{40}{840}$	$\frac{100}{840}$	$\frac{20}{84}$	$\frac{35}{84}$	$\frac{56}{84}$	$\frac{84}{84}$

Table 8-2. Increasing the weights of sub-pools 4 and 5

The procedure used to construct Table 8-2 is entirely natural and intuitive: begin by evaluating the case where each sub-pool appears without replication as illustrated in Table 8-1; then adjust each value in row 2 and row 3 by multiplying by the number of desired duplications for the corresponding column. Finally, use the adjusted values in rows 2 and 3 to compute the values in row 4.

This replication process is, of course, based on the implicit assumption that the relative weights of each sub-pool can be expressed as a set of integers. This assumption should be sufficient for most problems of practical interest.

For the example illustrated in Table 8-2, the total number of samples in row 3 is:

$$1+40+100+20+35+56+84 = 336$$

The total number of samples for which $n > 4$ is:

$$100+20+35+56+84 = 295$$

Thus, assuming that the sub-pools corresponding to $n = 4$ and $n=5$ are ten times more likely than the others, the chance that candidate A will win drops about 10% from 205/210 to 295/366 (i.e., from .976 to .878).

The replication of sub-pools makes it possible to employ Poincaré's assumption of equally probable cases in situations where all cases (i.e., all sub-pools) are not equally probable in the initial analysis. The replication process is similar in spirit to the approach Bayesians follow when they employ prior distributions that are not uniform.

8.4.4 Analysis of different sample

To make the example in the preceding section more interesting, suppose the three voters who were initially sampled did not all plan to vote for candidate A. Suppose instead that two voters were in favor of candidate A while the third supported candidate X. The analysis of this case proceeds exactly as before. The only change is in the formula used to compute the values in row 3 of Tables 8-1 and 8-2.

The values in row 3 represent the number of distinct samples that have the same composition as the observed sample. This observed sample now contains exactly two voters who favor candidate A. The third voter favors candidate X. If n is again equal to the number of voters who favor candidate A in the entire population, the number of samples with exactly two A voters is:

$$\frac{n\times(n-1)}{1\times2}\times(9-n)=n\times(n-1)\times(9-n)/2 \qquad (8\text{-}12)$$

Equation (8-12) replaces equation (8-11) for the computation of the values in row 3 of Table 8-1. The new values are shown in Table 8-3.

0	1	2	3	4	5	6	7	8	9
84	84	84	84	84	84	84	84	84	84
0	0	7	18	30	40	45	42	28	0
$\frac{0}{84}$	$\frac{0}{84}$	$\frac{7}{84}$	$\frac{18}{84}$	$\frac{30}{84}$	$\frac{40}{84}$	$\frac{45}{84}$	$\frac{42}{84}$	$\frac{28}{84}$	$\frac{0}{84}$

Table 8-3. Chance that exactly two voters in the sample favor candidate A

Thus, the chance that candidate A will win, under the assumption that all margins of victory are equally likely, is:

$$\frac{40+45+42+28}{7+18+30+40+45+42+28}=\frac{155}{210}=.738 \qquad (8\text{-}13)$$

If elections that are decided by a single vote (4 to 5 or 5 to 4) are ten times more likely than any alternative, the values in row 4 of Table 8-3 become: 0, 0, 7, 18, 300, 400, 45, 42, 28, 0. In this case, the chance that candidate A will win drops to:

$$\frac{400+45+42+28}{7+18+300+400+45+42+28}=\frac{515}{840}=.613 \qquad (8\text{-}14)$$

The reduced chance of a victory for candidate A in equation (8-14) is due to the fact that the chance candidate A will receive only 4 votes (and thus loose the election) has increased from 30/210 to 300/840 (i.e., from .143 to .357).

8.4.5 Traditional solution using Bayes' Theorem

The analysis presented in the preceding section closely parallels the traditional Bayesian approach to problems of this type. This section introduces some Bayesian concepts and terminology that relate directly to the election problem. Readers who are primarily interested in the further development of the proportionalist perspective rather than the step-by-step details of the corresponding Bayesian analysis can proceed directly to Section 8.6.

Bayes' Theorem is applied in cases where an initial (i.e., prior) probability distribution is being revised to reflect the availability of new information (i.e., evidence). The revision results in a new (i.e., posterior) distribution.

In the election example described in the previous section, the number of voters who favor candidate A is unknown. This unknown quantity is characterized by a prior distribution. Adapting the notation and terminology of (Kobayashi et al. 2012), let:

$P[A(n)]$ = prior probability that a total of n voters in the population favor candidate A (for $n = 0, 1, 2 \dots 9$)

Now suppose that a sample of three voters is drawn at random, and all three voters favor candidate A. Denote this event by the letter B. The objective of the analysis is to re-evaluate $P[A(n)]$, given that event B has been observed. More precisely, the objective is to evaluate the following probabilities:

$P[A(n)|B]$ = posterior probability that a total of n voters in the population favor candidate A (for $n = 0, 1, 2 \dots 9$), given that all three voters in the sample favor this candidate

Bayes' Theorem provides a formula for deriving the values of $P[A(n)|B]$. The formula depends on the values of $P[A(n)]$, the prior distribution. This formula also depends on the following set of conditional probabilities:

P[B|A(n)] = probability that all three voters in a sample favor candidate A, given that a total of n voters favor this candidate

Assuming the values of P[A(n)] and P[B|A(n)] are known for n = 0, 1, 2, ... 9, Bayes' Theorem states that each value of P[A(n)|B] can be computed as follows:

$$P[A(n)|B] = \frac{P[B|A(n)] \times P[A(n)]}{\sum_{j=0}^{9} P[B|A(j)] \times P[A(j)]} \tag{8-16}$$

As noted previously, the number of distinct three-voter samples that can be drawn from a population of nine voters is $(9 \times 8 \times 7)/(1 \times 2 \times 3) = 84$. Each of these 84 samples can be tagged with an integer from 0 to 9 to indicate the number of voters who favor candidate A in the population from which the sample is drawn. The resulting 840 possibilities represent the sample space for this application of Bayes' Theorem.

Following the discussion in the preceding sections, each member of this sample space is represented by a string of three letters followed by a numerical tag. The letters, which always appear in alphabetical order, range from A though I. The numerical tag, which ranges from 0 to 9, can be used to organize the 840 samples into ten sub-pools, each containing 84 samples that share a common tag.

P[A(n)] represents the probability that n voters favor candidate A. This is equivalent to stating that P[A(n)] is the probability that a sample selected at random will be drawn from sub-pool n. Since all 84 samples within each sub-pool have the same chance of being selected, the chance of selecting any individual sample within sub-pool n is P[A(n)]/84.

Event B is a subset of the sample space. It is comprised of those samples for which all three voters favor candidate A. As already noted, for each value of n, sub-pool n contains $n \times (n-1) \times (n-2)/6$ samples that are members of B. These are the values that appear in row 3 of Table 8-1.

P[B|A(n)] represents the probability that all three voters in a sample favor candidate A, given that the sample has been drawn from sub-pool n. These probabilities are equal to the proportions that appear in row 4 of Table 8-1.

The objective of the analysis is to evaluate P[A(n)|B]: the probability that a total of n voters favor candidate A, given that a sample comprised of three voters who favor candidate A has been drawn. The assumption that all three voters in the sample are members of B implies that many of the 840 members of the original sample space can be ignored. Only those samples that are members of B are now relevant.

Each of these samples has an unconditional probability of being selected: for a sample in sub-pool n, that probability is P[A(n)]/84. Since the number of samples in sub-pool n that are members of B is equal to $n \times (n-1) \times (n-2)/6$, the unconditional probability that a sample selected at random is a member of sub-pool n and also a member of B is equal to

$$\frac{n \times (n-1) \times (n-2)}{6} \times \frac{P[A(n)]}{84} = P[B \mid A(n)] \times P[A(n)]$$

Note that this term appears as the numerator on the right hand side of equation (8-16). The denominator simply renormalizes the ten values of $P[B \mid A(n)] \times P[A(n)]$ so they become a set of conditional probabilities whose sum is equal to 1. To evaluate equation (8-16) and solve the original problem, it is necessary to know the actual values of P[B|A(n)] and P[A(n)] for $n = 0, 1, 2 \ldots 9$.

As already noted, the last row of Table 8-1 contains the values of P[B|A(n)]. However, nothing in the statement of the original problem provides any information about the values of P[A(n)].

When no information about a prior distribution is available, it is common to assume that the distribution is uniform: in this case, to assume that all ten values of P[A(n)] are the same (i.e., equal to 1/10). When combined with the values that appear in the last row of Table 8-1, this assumption of a uniform prior yields the same solution presented in Section 8.4.2: the probability that candidate A will win is equal to 205/210.

Bayes' Theorem makes it simple to account for cases where the prior distribution is not uniform. For example, in the case where one-vote margins are ten times more likely than other possible outcomes, the values of P[A(4)] and P[A(5)] can be set to 10/28, and the remaining eight values of P[A(j)] to 1/28. Equation (8-16) then yields the same solution presented in Section 8.4.3. The probability that candidate A will win is equal to 295/336. There is no need to replicate sub-pools as in the case of Table 8-2.

8.5 Bayes' Theorem and the medical test

Return now to the example discussed in Section 8.2. An individual takes a medical test that indicates he or she has a disease. The accuracy of the test is quite good, but not perfect. Of those individuals who have the disease, 99% get a positive result. However, 1% of the individuals who do not have the disease get a false positive. The problem is to determine the chance the individual has the disease.

Interpreting chance as probability, the Bayesian analysis proceeds as follows:

A(0)   =   The set of individuals in the population who have the disease

A(1) = The set of individuals in the population who do not have the disease

B = The set of individuals who receive a positive test result

P[B|A(0)] = probability an individual's test result is positive, given that he or she has the disease

= .99

P[B|A(1)] = probability an individual's test result is positive, given that he or she does not have the disease

= .01

Table 8-4 is the analog of Table 8-1. In this case, the proportions in the last row again represent P[B|A(0)] and P[B|A(1)]. However, these conditional probabilities are not determined by a combinatoric analysis. They are instead the result of direct observation of the entire population or, perhaps, of a sample drawn from the population.

	Have the disease	Don't have the disease
Number tested	?	?
Number of positives	?	?
Proportion of positives	.99	.01

Table 8-4. Values needed for a Bayesian analysis

The prior distribution is also the result of direct observation. According to the statement of the problem, the proportion of individuals in the entire population who actually have the disease is equal to .01. Thus

P[A(0)] = prior probability an individual actually has the disease

$$= .01$$

P[A(1)] = prior probability an individual does not have the disease

$$= .99$$

The objective of the analysis is to determine the probability that an individual who tests positive actually has the disease. For this case, the general statement of Bayes' Theorem that appears in equation (8-16) can be simplified considerably as shown in equation (8-17).

P[A(0)|B] = posterior probability that an individual has the disease, given that the result of his or her medical test is positive

$$= \frac{P[B \mid A(0)] \times P[A(0)]}{P[B \mid A(0)] \times P[A(0)] + P[B \mid A(1)] \times P[A(1)]} \qquad (8\text{-}17)$$

The values of P[B|A(0)] and P[B|A(1)] appear in row 3 of Table 8-4, and the values of P[A(0)] and P[A(1)] are known to be .01 and .99 respectively. Substituting these values into equation (8-17) yields

$$P[A(0) \mid B] = \frac{.99 \times .01}{.99 \times .01 + .01 \times .99} = .50$$

Thus, the probability an individual has the disease, given that the result of his or her medical test is positive, is equal to one half. As already noted, this agrees with the proportionalist result derived in Section 8.2.2.

8.6 Finite and infinite populations

All the proportions that appear in the equations of this chapter are obtained by dividing finite numerators by finite denominators. An equally important but less obvious point is that each term in these numerators and denominators represents a quantity that can be evaluated by measuring a finite number of observable elements.

These assumptions satisfy the requirements of most practitioners since the number of observable and measurable quantities that practitioners deal with is invariably finite. A number of prominent theoreticians have recognized this point. Kolmogorov's comment (1933) that "any observable random process" can be characterized by a "finite field of probability" has already been cited in Section 1.2.7. In addition, J.D.C. Little, whose work is discussed in Section 7.9, has written (2011): "Whereas the usual proofs [of Little's Law] assume that the process runs over a time axis $(0 \leq t < \infty)$, practitioners operate over a finite time period [0,T]."

Nevertheless, mathematicians have an entirely legitimate interest in questions that lie beyond these finite boundaries. To cite a particularly well known example, suppose a number is drawn at random from the interval [0,1]: what is the chance that this number is rational?

From a proportionalist perspective, the answer to this question is obtained by determining the total amount of space in the interval [0,1] that is occupied by rational numbers and then dividing this value by 1, the length of the entire interval. Both quantities are clearly finite. Mathematicians refer to the numerator as the measure of the rational numbers in [0,1] and the denominator as the measure of the entire interval [0,1].

The straightforward measurement procedures described in Section 3.7 cannot be used to evaluate the numerator because the number of rational numbers in [0,1] is infinite and the amount of space occupied by each individual rational number is zero. This is also true of the irrational numbers in [0,1]. The measure of the entire interval [0,1] is clearly equal to 1, but determining the measure of the rational and irrational subsets that combine to form this interval is an exquisitely subtle problem.

The branch of modern mathematics that deals with questions of this type is known as measure theory. Section 9.2.7 includes a very brief introduction to this topic and describes the standard measure theoretic procedure for determining the measure of the set of rational numbers in [0,1].

Measure theory is based on a few simple and highly intuitive axioms that define the formal mathematical properties of functions known as measures. The measurement procedures described in Section 3.7 conform to these axioms. Thus these procedures define a legitimate mathematical measure. However, the scope of this measure is limited to sets comprised of a finite number of observable elements. These elements include discrete events that can be counted using pattern matching algorithms and continuous intervals whose durations can be measured by subtracting starting times from ending times. Geometric shapes in two or more dimensions can also be accommodated, provided the areas or volumes bounded by these shapes can be computed using standard geometric formulas or ordinary Riemann integration.

As discussed in Section 9.2.7, modern measure theory is substantially broader in scope. It incorporates a generalized approach to measurement devised by Émile Borel and extended by Henri Lebesgue. Despite the underlying power and mathematical elegance of this approach, all subsets of [0,1] are not necessarily Lebesgue measurable. In fact, it is possible to prove that the interval [0,1] contains an infinite number of such non-measurable subsets. [Note: this proof requires the somewhat controversial Axiom of Choice.]

Viewed from this perspective, the straightforward definitions of measurement described in Section 3.7 can be regarded as defining a simple "practitioner's measure" that generates the same result as Borel measure or Lebesgue measure in situations where all three measures yield well defined values. For the majority of practitioners who deal with problems that arise in the physical world rather than the realm of abstract mathematics, these are the only cases that really matter.

CHAPTER 9

Limiting Properties of Samples

9.1 The Law of Large Numbers

The Law of Large Numbers establishes a linkage between the observable properties of random samples and the properties of underlying populations from which such samples are drawn. The next few sections provide a proportionalist interpretation of this well known and fundamentally important result.

As usual, it is helpful to begin with a simple example. Suppose an urn contains a number of colored balls. An experimenter reaches into the urn, selects a ball at random, notes its color and then returns the ball to the urn. Assume this basic sampling procedure is repeated for a total of N cycles.

Let k = total number of balls in the urn

g = number of balls in the urn that are green

N = number of sampling cycles (i.e., number of times a ball is drawn from the urn and then replaced)

If N is large, it is reasonable to expect that the fraction of green balls in the N samples should be close to the fraction of green balls in the urn itself. Since g/k is the fraction of green balls in the urn, it follows that the total number of green balls drawn during the experiment should be approximately equal to $N \times (g / k)$.

This result cannot be guaranteed with certainty because of the random nature of the sampling process. However, its validity should become increasingly likely as the value of N grows larger. The Law of Large Numbers provides a mathematically rigorous characterization and proof of this intuitively appealing notion.

9.1.1 Basic proportionalist relationships

To discuss the Law of Large Numbers in proportionalist terms, begin by assuming that the balls in the urn are numbered 1 through k. Each sequence of N successive samples can then be represented by a sequence of N integers, where each integer lies in the range from 1 to k. The same ball can be drawn multiple times, so individual integers may appear in these sequences more than once. The total number of distinct numerical sequences that can be obtained when N samples are drawn from an urn containing k balls is clearly equal to k^N.

From a proportionalist perspective, the notion that each ball has been drawn at random implies that each of these k^N numerical sequences has the same chance of being observed (i.e., all k^N numerical sequences are equally probable in Poincaré's terminology). This characterization of chance, which follows directly from the discussion in Chapter 8, is the key to the proportionalist formulation of the Law of Large Numbers.

Let G = number of green balls that appear in the sample

The next step is to determine the number of distinct sequences that contain exactly G green balls (G is any value from 0 to N). Begin by considering those sequences in which the first G samples are all green and the following N-G samples are all of some other color. The number of distinct sequences with this particular arrangement of green and non-green balls is clearly equal to $g^G (k - g)^{N-G}$.

There are, of course, many other ways to arrange sequences of N samples so that exactly G green balls are drawn. Standard combinatorial arguments imply that the number of different ways G green balls and N-G non-green balls can be arranged in a sequence of length N is equal to

$$\frac{N!}{G!(N-G)!} = \binom{N}{G} \tag{9-1}$$

Each of these permutations of G green balls and N-G non-green balls can be associated with $g^G \times (k-g)^{N-G}$ distinct numerical sequences. Thus, the total number of distinct numerical sequences of length N that contain exactly G green balls is

$$\binom{N}{G} g^G (k-g)^{N-G} \tag{9-2}$$

Since the total number of distinct sequences of length N is equal to k^N, the proportion of these sequences that contain exactly G green balls is

$$P(N,G,g,k) = \binom{N}{G} \frac{g^G (k-g)^{N-G}}{k^N}$$

$$= \binom{N}{G} (g/k)^G (1-g/k)^{N-G} \tag{9-3}$$

Equation (9-3) is valid for G = 0, 1, 2 ...N. For each value of G, equation (9-3) specifies the proportion of sequences of length N that contain exactly G green balls, assuming these sequences are generated by drawing balls from an urn containing k balls, where g of these balls are green. If it is further assumed that the drawings are made at random, all possible sequences are equally probable (in the sense of Poincaré and Section 8.2.3), and the proportions specified by equation (9-3) can be reinterpreted as probabilities. Under this reinterpretation, equation (9-3) becomes the specification of a traditional binomial distribution with parameters N, G and g/k.

9.1.2 Limiting behavior

As already noted, it is reasonable to expect the value of G to be close to $N \times (g / k)$ if N is large. Before formulating this notion in precise mathematical terms, it is worthwhile to examine some specific numerical examples. Begin with the case where $g / k = \frac{1}{2}$, which implies that G is likely to be close to N/2 when N is large. If N is equal to 10, the value of P(10, 5, 1/2) is 24.6% as shown in equation (9-4).

$$P(10,5,\tfrac{1}{2}) = \frac{252}{1024} = 24.6\% \qquad (9\text{-}4)$$

If N is increased to 20, the corresponding value of P(20, 10, ½) becomes 17.6% as shown in equation (9-5).

$$P(20,10,\tfrac{1}{2}) = \frac{184,756}{1,048,576} = 17.6\% \qquad (9\text{-}5)$$

This result may, at first, appear counter-intuitive. Increasing the value of N from 10 to 20 has actually reduced the chance that G is exactly equal to N/2. This is due to the fact that the number of possible values for G has increased from 11 to 21, making it less likely that G will be exactly equal to any given value.

A better way to look at this issue is to consider the chance that the value of G falls into some range that is centered at N/2: for example $0.4 \le G/N \le 0.6$. For N =10, this range includes G=4 and G=6 as well as G=5.

$$P(10,4,\tfrac{1}{2}) = \frac{210}{1024} = 20.5\% \qquad (9\text{-}6)$$

$$P(10,6,\tfrac{1}{2}) = \frac{210}{1024} = 20.5\% \qquad (9\text{-}7)$$

Combining equations (9-4), (9-6) and (9-7) implies that the chance that G/N lies between 0.4 and 0.6 is equal to:

$$24.6\% + 20.5\% + 20.5\% = 65.6\%.$$

The same calculation can now be performed for the case where N=20. There are now five possible values of G that correspond to values of G/N between 0.4 and 0.6: 8, 9, 10, 11, and 12. Note that

$$P(20,9,\tfrac{1}{2}) = \frac{167,960}{1,048,576} = 16.0\% \qquad (9\text{-}8)$$

$$P(20,11,\tfrac{1}{2}) = \frac{167,960}{1,048,576} = 16.0\% \qquad (9\text{-}9)$$

$$P(20,8,\tfrac{1}{2}) = \frac{125,970}{1,048,576} = 12.0\% \qquad (9\text{-}10)$$

$$P(20,12,\tfrac{1}{2}) = \frac{125,970}{1,048,576} = 12.0\% \qquad (9\text{-}11)$$

Combining equations (9-8) through (9-11) with equation (9-5) implies that the chance that G/N lies between 0.4 and 0.6 is:

$$17.7\% + 16.0\% + 16.0\% + 12.0\% + 12.0\% = 73.8\%.$$

The chance that G/N is between 0.4 and 0.6 will continue to increase steadily each time N is doubled.

Conversely, the chance that G/N will fall outside the range [0.4, 0.6] decreases steadily each time N is doubled: this chance is $100\% - 65.6\% = 34.4\%$ when $N = 10$ and drops to 26.2% when $N = 20$. When N grows to 80, the chance that G/N will fall outside the range [0.4, 0.6] drops all the way to 5.6% as shown in Table 9-1. If additional rows are added to this table to represent larger values of N, the chance that G/N is outside the range [0.4, 0.6] will continue to drop – more or less steadily – and eventually approach zero. In precise mathematical terms, for any $\varepsilon > 0$ it is always possible to find a value of NMIN such that the values in this column are less than ε for all values of N greater than NMIN.

The next column in Table 9-1 shows the impact of tightening the desired range from [0.40, 0.60] to [0.45, 0.55]. For any given value of N, the corresponding chance that G/N will fall outside this tighter range is of course larger. Nevertheless, the values in this column also approach zero as N increases. Once again, it is always possible to find a value of NMIN such that the values in this column are less than ε for all values of N greater than NMIN.

N	Chance $|G/N-g/k| > 0.10$	Chance $|G/N-g/k| > 0.05$
10	34.4%	75.4%
20	26.2%	50.3%
40	15.4%	43.0%
60	9.2%.	36.6%
80	5.6%	31.4%
100	3.6%	27.2%

Table 9-1. Chance $|G/N - g/k|$ exceeds thresholds when $g/k=1/2$

Now suppose another column is added to Table 9-1. Assume the desired range associated with this column is arbitrarily small and can be represented as $[g/k-\Delta, g/k+\Delta]$ for some small positive value of Δ. In other words, each value in this column represents the chance that the difference between G/N and g/k is greater than Δ for the associated value of N. Equation (9-3) can once again be used to compute the values for this column.

Regardless of the value of Δ, the chance that the difference between G/N and g/k is greater than Δ will always approach zero as N approaches infinity. In formal mathematical terms, for any $\Delta > 0$ and any $\varepsilon > 0$, it is always possible to find a value of NMIN such that the values in this column are less than ε for all values of N greater than NMIN.

This formal statement, which represents the essence of the Law of Large Numbers, is a consequence of equation (9-3). Intuitively, as N grows larger, the proportion of sequences for which G/N is close to g/k will increase and the proportion of sequences for which G/N is more than a fixed distance Δ from g/k will become vanishingly small. A rigorous proof of this intuitively reasonable notion is beyond the scope of this discussion.

9.1.3 Extension to complete distributions

The analysis in Sections 9.1.1 and 9.1.2 focused on the number of green balls drawn from the urn during a sequence of N random samples. The same analysis can clearly be applied to balls of any other color. For example, if the urn contains a mixture of balls with different colors as shown in Table 9-2 and a large number of samples are drawn at random, it is reasonable to expect that the proportion of red balls will be close to 1/16, the proportion of orange balls will be close to 2/16, and so on.

Formally, it is possible to show that for any $\Delta > 0$ and $\varepsilon > 0$, it is always possible to find a value of NMIN that satisfies the following property: consider the chance that the observed proportion of balls of any color is more than Δ from its corresponding proportion in Table 9-2; this chance is less than ε for all values of N greater than NMIN. In technical terms, the observed distribution of colors will converge uniformly to the distribution in the rightmost column in Table 9-2. This is another implication of the Law of Large Numbers.

Color	Wavelength (nm)	Number of Balls	Proportion
Red	700	1	1/16
Orange	620	2	2/16
Yellow	580	3	3/16
Green	530	5	5/16
Blue	470	3	3/16
Violet	420	2	2/16

Table 9-2. Distribution of balls in the urn

9.1.4 Traditional Law of Large Numbers

To complete the discussion of this example, note that Table 9-2 also lists the wavelength (in nanometers) for each of the six colors. This makes it possible to display the information in Table 9-2 using the graph in Figure 9-1. Note that the horizontal axis is wavelength in nanometers and the vertical axis represents the proportion of balls having each associated color.

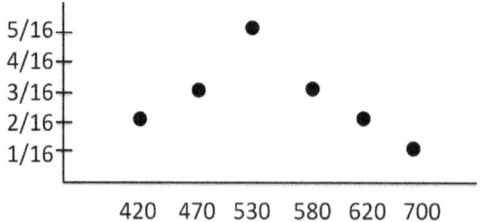

Figure 9-1. Proportion of balls of each color

It is a simple matter to compute the average wavelength for all 16 balls in the urn, where each wavelength is weighted by the number of balls of the corresponding color. This computation is shown in equation (9-12):

$$\frac{1\times700+2\times620+3\times580+5\times530+3\times470+2\times420}{16}=536.25$$

(9-12)

Imagine performing the same computation for a sample of N balls that are drawn at random from this urn. Since the proportion of time each color is observed is likely to be close to the corresponding proportion in Table 9-2, it follows that the observed average wavelength in the sample should be close to the underlying value 536.25 if N is large. In fact, the limit – as N approaches infinity – of the observed average wavelength is equal to the underlying expected value computed in equation (9-12). This limiting property of observed averages represents the traditional form of the Law of large Numbers.

In precise mathematical terms, the Law of Large Numbers states that, for any $\Delta>0$ and $\varepsilon>0$, it is always possible to find a value of NMIN with the following property: consider the chance that the observed average wavelength is more than Δ from the underlying expected value in equation (9-12); this chance is less than ε for all values of N greater than NMIN.

This form of the Law of Large Numbers also applies to continuous distributions of the type illustrated in Figure 9-2. Such distributions cannot be represented faithfully by assuming that balls are being drawn at random from an urn with a finite capacity. As a result, the simple weighted average in equation (9-12) must be replaced by an integral. Thus the simple and intuitively appealing arguments presented in Sections 9.1.2 and 9.1.3 are not directly applicable to the more general case of sampling from random variables with continuous probability distributions. Although this limitation is a legitimate concern for mathematicians, it is of less importance to most practitioners. The Epilog presents a further discussion of this issue.

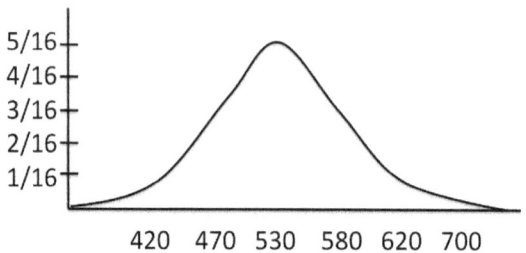

Figure 9-2. A continuous distribution of colors

9.2 The Ergodic Theorem

The Law of Large numbers deals with the relationship between the observable properties of large numbers of random samples and the properties of the underlying population from which these samples are drawn. The Ergodic Theorem also deals with relationships between samples and underlying populations, but in this case the samples are not statistically independent and the populations have complex structures that cannot be represented as urns containing finite numbers of balls of various colors.

Despite these differences, the Ergodic Theorem and the Law of Large Numbers both lead to conclusions that make good sense on an intuitive level: they both imply that important observable properties of samples will, in the limit, approach the corresponding properties of the underlying population. However, these con-clusions are substantially more difficult to characterize rigorously in the case of the Ergodic Theorem because the term "in the limit" cannot be interpreted in the conventional manner. The next few sections examine the reasons these difficulties arise and the formalisms that have been devised to overcome them.

9.2.1 Proportionalist analog of a simple stochastic process

The simple random walk discussed in Chapters 1, 2 and 3 provides a useful starting point for this discussion. As usual, assume the walker travels between four stations as illustrated in Figures 1-3 and 2-1. Suppose the direction the walker turns after leaving each station is determined by selecting a ball at random from an urn. If a red ball is selected, the walker turns to the right; if a white ball is selected, the walker turns to the left. When the walker enters the next station, the selected ball is returned to the urn and another random sample is drawn to determine the direction of the next turn. To make this example as simple as possible, assume the urn contains an equal number of red and white balls.

Table 9-3 illustrates all possible trajectories of length less than or equal to four.

The first column contains the four possible initial stations from which the walk can begin: 0, 1, 2 and 3.

The second column contains the eight possible trajectories that can be generated after one turn: 00, 01, 10, 12, 21, 23, 32, 33.

The third column contains the sixteen possible trajectories that can be generated after two turns: 000, 001, 010, 012, 100, 101, 121, 123, 210, 212, 232, 233, 321, 323, 332, 333.

The fourth and fifth columns contain the thirty-two possible trajectories that can be generated after three turns: 0000, 0001, 0010, 0012, 0100, 0101, 0121, 0123, 1000, 1001, 1010, 1012, 1210, 1212, 1232, 1233, 2100, 2101, 2121, 2123, 2321, 2323, 2332, 2333, 3210, 3212, 3232, 3233, 3321, 3323, 3332, 3333.

In general, the number of possible trajectories that can be generated after N turns is 4×2^{N}.

Length of Trajectories				
1	2	3	4	4
0	00	000	0000	2100
1	01	001	0001	2101
2	10	010	0010	2121
3	12	012	0012	2123
	21	100	0100	2321
	23	101	0101	2323
	32	121	0121	2332
	33	123	0123	2333
		210	1000	3210
		212	1001	3212
		232	1010	3232
		233	1012	3233
		321	1210	3321
		323	1212	3323
		332	1232	3332
		333	1233	3333

Table 9-3. Possible Trajectories of length 1, 2, 3 and 4

9.2.2 Transient and limiting properties

Since there are an equal number of red and white balls in the urn, all the trajectories that begin from the same initial station have the same chance of being followed. Thus, if the walker begins at station 0, the first two trajectories in column 2 have the same chance (i.e., ½) of being followed, the first four trajectories in column 3 have the same chance (i.e., ¼) of being followed, and so on.

In proportionalist terminology, the chance that a walker starting at station 0 is at station 0 after the first turn is equal to ½, the chance that this walker is at station 1 after the first turn is also equal to ½, and the chance that this walker is at station 2 or 3 after the first turn is 0. The same analysis implies that the chances this walker is at stations 0, 1, 2 or 3 after the second turn are ½, ¼, ¼ and 0 respectively. After the third turn, these chances become $^3/_8$, $^3/_8$, $^1/_8$ and $^1/_8$. These chances, which can be calculated for each successive value of N, correspond to traditional transient probabilities.

This method for dealing with transient probability distributions in observational stochastics is clearly applicable to any model of the type discussed in Chapter 4. In this particular case, Table 9-4 illustrates how these values change as the length of the trajectory increases. Not surprisingly, these values approach stable limits: each column contains a sequence of values that converges to 0.25. This convergence is similar to the behavior illustrated in Table 2-1 and is consistent with equations (2-21) through (2-24) with r = ½. Intuitively, the proportions 0.25, 0.25, 0.25 and 0.25 represent the chance that a walker starting at station 0 is at station 0, 1, 2 or 3 after thirteen or more steps. These proportions are of course analogous to traditional steady state probabilities.

N=Number of Turns (steps)	Proportion of trajectories that end with the walker at:			
	Station 0	Station 1	Station 2	Station 3
0	1.000	0.000	0.000	0.000
1	0.500	0.500	0.000	0.000
2	0.500	0.250	0.250	0.000
3	0.375	0.375	0.125	0.125
4	0.375	0.250	0.250	0.125
5	0.313	0.313	0.187	0.187
6	0.313	0.250	0.250	0.187
7	0.281	0.281	0.219	0.219
8	0.281	0.250	0.250	0.219
9	0.266	0.266	0.234	0.234
10	0.266	0.250	0.250	0.234
11	0.258	0.258	0.242	0.242
12	0.258	0.250	0.250	0.242
13	0.254	0.254	0.246	0.246

Table 9-4. Position of the walker after N turns

9.2.3 Trajectory-based proportions

The analysis in the preceding section has focused on the final location of the walker after a fixed number of turns, assuming the walker has an equal chance of following a variety of different trajectories. It is also possible to consider the behavior of the walker while moving along a single trajectory from its initial station (e.g., station 0) to its final station. The number of visits the walker makes to each station in that particular trajectory can then be computed.

For each value of N and each initial state, the number of possible trajectories a walker can follow is equal to 2^N. Column 1 of Table 9-5 displays the eight possible trajectories for the case where the walker begins at station 0 and then makes three turns (i.e., N=3) in succession. The remaining columns of Table 9-5 present the proportion of visits the walker makes to each station over the course of that trajectory.

Specific Trajectory	Proportion of visits to each station			
	Station 0	Station 1	Station 2	Station 3
0000	1.00	0.00	0.00	0.00
0001	0.75	0.25	0.00	0.00
0010	0.75	0.25	0.00	0.00
0012	0.50	0.25	0.25	0.00
0100	0.75	0.25	0.00	0.00
0101	0.50	0.50	0.00	0.00
0121	0.25	0.50	0.25	0.00
0123	0.25	0.25	0.25	0.25

Table 9-5. Proportion of visits for specific trajectories

9.2.4 Implications of the Ergodic Theorem

The Ergodic Theorem is concerned with trajectory-based proportions of the type shown in each row of Table 9-5. The proportion of visits to each station varies substantially from row to row in this simple case where $N = 3$. However, as N increases, the proportions that appear in each column of the vast majority of rows will all be very close to .25. In other words, most rows in the extended version of Table 9-5 will contain values that are approximately equal to the proportions that appear in the final row of the extended version of Table 9-4.

As already noted, the values that appear in the final row of Table 9-4 correspond, in the limit, to steady state probabilities. In contrast, the values that appear in each row of Table 9-5 correspond to observed proportions for individual trajectories. The fact that these observed proportions are close to the corresponding steady state probabilities for the vast majority of trajectories is the essence of the Ergodic Theorem.

The Ergodic Theorem is critically important to practitioners. When a practitioner validates a model by observing a real system over some interval of time or running a Monte Carlo simulation, the practitioner is working directly with one, and only one, trajectory. This corresponds to exactly one row of Table 9-5. On the other hand, traditional stochastic models deal with steady state probability distributions that correspond to the final row of Table 9-4. Without a way to link these two conceptually distinct quantities, the practical value of traditional stochastic models would be severely limited.

The formal proof of the Ergodic Theorem is complicated by the fact that it is valid for most – but not all – trajectories (i.e., most, but not all, rows in Table 9-5). [This is so even when trajectory length becomes infinite.] The main technical challenge is to characterize the relatively rare cases in which the theorem is incorrect. This highly technical issue is examined in Section 9.2.7.

9.2.5 Uncertain initial states

The discussion in the preceding two sections is based on the assumption that the walker begins at station 0. As already noted, similar analyses can be carried out if the walker begins at stations 1, 2 or 3. Suppose, however, that the walker's starting location is determined by drawing two balls at random from an urn with an equal number of red and white balls. As usual, assume the first ball is placed back in the urn before the second sample is drawn. If two successive samples are red, the walker starts from station 0. If both samples are white, the walker starts from station 1. If the first sample is red and the second is white, the walker starts from station 2. Finally, if the first sample is white and the second is red, the walker starts from station 3.

Assuming that each drawing is made at random, it is appropriate to assume that each station has a 25% chance of being the starting station for the walk. In other words, all four trajectories in column 1 of Table 9-3 have an equal chance of being followed. This then implies that all eight trajectories in column 2 of Table 9-3 have an equal chance of being followed, and so on for all possible trajectories of any specific length.

This example is especially interesting because the chances the walker is at stations 0, 1, 2 and 3 are all equal to 0.25 after each successive step. These chances are the same as the steady state probability distribution obtained by evaluating equations (2- 21) through (2-24) with $r = \frac{1}{2}$. In other words, the stochastic process associated with this case actually begins in steady state and thus remains in steady state forever. This is another example of the principle illustrated in Table 2-4.

9.2.6   Cases where r ≠ ½

The proportionalist arguments presented in the preceding sections depend on the assumption that all trajectories originating from a given initial station have the same chance of being followed. This assumption is intuitively reasonable when the number of red balls in the urn is equal to the number of white balls (i.e., r= ½), and when every ball in the urn has the same chance of being drawn (i.e., when balls are drawn from the urn at random).   Such cases clearly satisfy Poincaré's requirement that all trajectories be equally probable.

This assumption can be relaxed by assuming that the number of red balls in the urn differs from the number of white balls. In such cases the chance the walker follows trajectory 00 is different from the chance the walker follows trajectory 01. These differences can be accommodated by replicating the number of times trajectories 00 and 01 appear in the revised version of Table 9-3.

For example, if the urn contains one red ball and two white balls, the chance the walker turns left when exiting from station 0 is twice the chance the walker turns right. This can be represented by modifying Table 9-3 so that trajectories generated when the walker turns to the left appear twice as frequently as trajectories generated when the walker turns to the right. Table 9-6 presents an example of this modification for N=2 and N=3.

Note that the number of trajectories in the second column has increased from eight to twelve. Similarly, the sixteen trajectories in the third column of Table 9-3 have increased to the thirty-six trajectories that appear in the remaining four columns of Table 9-6.

Replications of this type make it possible to define the chance that a particular trajectory is followed as the proportion of times that trajectory appears in the appropriate column(s) of Table 9-6.   This

Length of Trajectories					
1	2	3	3	3	3
0	00	000	100	210	321
1	00	000	100	210	321
2	01	001	101	212	323
3	10	000	100	210	321
	10	000	100	210	321
	12	001	101	212	323
	21	010	121	232	332
	21	010	121	232	332
	23	012	123	233	333
	32				
	32				
	33				

Table 9-6. Modified version of Table 9-3 for r = 1/3

computationally cumbersome but intuitively meaningful strategy mirrors the strategy described in Section 8.4.3. It can be extended to urns containing arbitrary numbers of red and white balls. In such cases, the value of r is simply the number of red balls divided by the total number of balls in the urn. If equations (2-21) through

(2-24) are then solved using this value of r, the resulting steady state distribution will yield the correct limiting distribution – as N approaches infinity – for the last row of the new version of Table 9-4 and for almost all rows of the new version of Table 9-6.

9.2.7 Role of measure theory

The Ergodic Theorem deals with the properties of ergodic stochastic processes. Each of these processes has a steady state distribution and is associated with an ensemble (i.e., a set) of associated trajectories. Essentially, the Ergodic theorem states that trajectory-based proportions will, in almost all cases, converge to the steady state distribution of the associated stochastic process as the length of the trajectory approaches infinity.

Note that convergence is not guaranteed. Some trajectories associated with an ergodic stochastic process may fail to converge. The main technical challenge in the proof of the Ergodic Theorem is to show that these non-convergent trajectories represent a negligible fraction of the entire ensemble and can be safely ignored. Measure theory provides a set of tools and formal concepts that can be used to demonstrate this point.

The key to the proof of the Ergodic Theorem is the concept of Borel measure, developed by Émile Borel to characterize the total length (i.e., the measure) of any given subset of the real line. In general, each subset is comprised of segments and individual discrete points. If the number of segments and points is finite, Borel's method and the straightforward measurement procedure described in Section 3.7.4 yield exactly the same results. However, Borel's definition also accommodates difficult cases where the number of segments and/or the number of points may be infinite. Under Borel's characterization, the measure of such a subset can be zero even though it contains an infinite number of members.

George Birkhoff's celebrated proof of the Ergodic Theorem [Birkhoff 1931] is based on an unexpectedly powerful application of the concept of Borel measure. Specifically, Birkhoff demonstrated that the set of non-convergent trajectories (i.e., trajectories whose observable proportions do not converge to the steady state distribution of the associated stochastic process) has Borel measure zero. Thus, in order to appreciate the strengths and limitations of the Ergodic Theorem, it is important to understand what it means for a set to have measure zero.

Essentially, the Borel measure of any set of discrete points or continuous intervals is the combined length of the smallest set of non-overlapping intervals that cover (i.e., contain) these points or intervals completely. This intuitively appealing notion of covering the set being measured by another set of intervals and then shrinking the lengths of the covering intervals to their minimum size is the central principle behind Borel's notion of measurement.

The length of each interval in the covering set is computed, as described in Section 3.7.4, by subtracting its starting position from its ending position. The combined length of all covering intervals is then computed by a summation that may involve an infinite number of terms. If the sum converges to a well-defined limit, that limiting value is the Borel measure of the original set.

If, for every $\varepsilon > 0$, it is possible to find a set of non-overlapping covering intervals whose combined length is less than or equal to ε, then the Borel measure of this set is equal to zero. This procedure for characterizing sets of measure zero is the crown jewel of Borel measure and a pillar of modern measure theory. In probability theory, which can be regarded as a special application of measure theory, an event is said to occur with *probability zero* if its Borel measure is zero. Similarly, and event is said to occur *almost surely* if its Borel measure is equal to one (i.e., if its complement has measure zero).

One of the best known examples of a set of measure zero is the set of rational numbers in the interval [0,1]. Even though this set contains an infinite number of members, the set is still countable as defined by the nineteen century mathematician Georg Cantor. This means its members can be arranged to form a simple linear sequence as shown in equation (9-13).

Set of rational numbers in $[0,1] =$

$$\left\{ 0,1,{}^1\!/_2, {}^1\!/_3, {}^2\!/_3, {}^1\!/_4, {}^3\!/_4, {}^1\!/_5, {}^2\!/_5, {}^3\!/_5, {}^4\!/_5 , {}^1\!/_6, {}^5\!/_6, \ \dots \right\} \qquad (9\text{-}13)$$

The first member of this sequence can be covered by an interval of length $\varepsilon/2$, the second by an interval of $\varepsilon/2^2$, the third by an interval of length $\varepsilon/2^3$, and so on. The sum of the lengths of this infinite sequence of partially overlapping covering intervals, which clearly represents an upper bound on the set's Borel measure, is equal to ε as shown in equation (9-14). Since ε can be made arbitrarily small, the Borel measure of the set of rational numbers in [0,1] must thus be equal to zero.

$$\varepsilon = \sum_{n=1}^{\infty} \frac{\varepsilon}{2^n} \qquad (9\text{-}14)$$

In the terminology of probability theory, if a number is drawn at random from the set of real numbers in the interval [0,1], it will almost surely be irrational. Drawing a rational number under these circumstances represents an event that occurs with probability zero. Although these conclusions are consistent with the mathematically precise definition of Borel measure, they may seem counter-intuitive to practitioners since the interval [0,1] actually contains an infinite number of rational numbers. Measure theory has many other implications that are also difficult to reconcile on an intuitive level (Gelbaum and Olmsted 1964).

A more exotic example of a set of measure zero is the Cantor set, also known as the Cantor ternary set. This set is constructed by systematically removing the open middle third of every closed interval in [0,1] *ad infinitum*. The first three removal cycles are depicted in Table 9-7. It is easy to show that the measure of the Cantor set is zero. However, this set is still fundamentally larger than the set of rational numbers. Technically, the Cantor set has the cardinality of the continuum and contains an uncountably infinite number of members, while the set of rational numbers in [0,1] has a lower cardinality (i.e., $\aleph_0$) and is merely countably infinite.

The existence of the Cantor set raises further concerns regarding the Ergodic Theorem. The fact that this theorem is almost surely valid for the trajectories in a given ensemble does not prevent the number of non-convergent trajectories within that ensemble from being uncountably infinite.

The Cantor set also has certain exceptional properties that are of special interest to mathematical theorists. For example, since the Borel measure of the Cantor set is zero, it is reasonable to expect that all subsets of the Cantor set must also have measure zero. However, this is not so: some subsets of the Cantor set are not Borel measurable.

Borel's student Henri Lebesgue refined Borel's original definition of measure to ensure that all subsets contained within a set of measure zero are also measurable and also have measure zero. This refinement is known as Lebesgue measure. In technical terms, Lebesgue measure represents the completion of Borel measure.

[0,1]	Initial interval
	Remove middle third from initial interval
[0, $^1/_3$]	After first removal
[$^2/_3$, 1]	After first removal
	Remove middle thirds from remaining intervals
[0, $^1/_9$]	After second set of removals
[$^2/_9$, $^3/_9$]	After second set of removals
[$^6/_9$, $^7/_9$]	After second set of removals
[$^8/_9$, 1]	After second set of removals
	Remove middle thirds from remaining intervals
[0, $^1/_{27}$]	After third set of removals
[$^2/_{27}$, $^3/_{27}$]	After third set of removals
[$^6/_{27}$, $^7/_{27}$]	After third set of removals
[$^8/_{27}$, $^9/_{27}$]	After third set of removals
[$^{18}/_{27}$, $^{19}/_{27}$]	After third set of removals
[$^{20}/_{27}$, $^{21}/_{27}$]	After third set of removals
[$^{24}/_{27}$, $^{25}/_{27}$]	After third set of removals
[$^{26}/_{27}$, 1]	After third set of removals

Table 9-7. Construction of the Cantor ternary set

Even after the process of completion, a modicum of untidiness remains. In particular, it is not true that all subsets of a Lebesgue measurable set are Lebesgue measurable. This intuitively appealing relationship is, in fact, only valid for sets of measure zero.

Lebesgue's work had a powerful influence on Kolmogorov, whose formulation of the axiomatic theory of probability "would have been a rather hopeless...task...before the introduction of Lebesgue's theories of measure and integration" (Kolmogorov 1933). As a result, many crucial results in modern probability theory are only valid for sets that are Lebesgue measurable. This constraint, which is required to rule out certain unusual counter-examples, has little if any practical impact: the finite sets of observations that practitioners deal with are always Lebesgue measureable.

Epilog

Observational stochastics deals directly with observable properties of individual trajectories whose lengths are finite. This makes it possible to analyze such properties without appealing to the Law of Large Numbers or the Ergodic Theorem. As a result, none of the formalities and subtle mathematical complexities discussed in Chapter 9 ever arise. There is no need to consider concepts such as Lebesgue measurability, sets of measure zero, and results that are almost surely, but not always, valid. The results derived through observational stochastics can be applied to directly observable quantities with complete certainty in all cases where the directly verifiable assumptions of the corresponding LCD model are satisfied.

Although these benefits are of genuine value to practitioners, the most important contributions of observational stochastics are conceptual rather than pragmatic. At its core, observational stochastics provides an alternative framework for thinking about and analyzing the behavior of systems in situations that involve uncertainty and randomness. In such cases, traditional stochastic analyses begin by invoking the sampling premise: that is, by assuming that observed values can be regarded as samples that have been drawn at random from underlying probability distributions.

The sampling premise shifts the focus of the analysis to probability distributions and their parameters. As pointed out at the start of Chapter 1, a model based on these entities is incompatible with the observability principle, which requires that all symbolic variables used in a mathematical model must correspond to directly observable and measurable properties of the system being modeled.

The exact meaning of *directly observable and measurable* is open to interpretation. However, it seems clear that this term is not applicable to quantities that can only be evaluated through procedures that require the Law of Large Numbers or the Ergodic Theorem for their formal justification.

The symbolic variables employed in observational stochastics are entirely different. By design, they are intended to represent quantities that are directly observable and measurable within the formal modeling framework developed in Section 3.7.

Observational stochastics must also provide a mechanism for representing uncertainty. This is achieved by regarding the directly observable and measurable values of uncertain quantities as immaterial details whose aggregate properties are subject to model-specific loose constraints. This alternative approach to the characterization of uncertainty sets observational stochastics apart from traditional stochastic modeling.

Note that the two modeling frameworks are not mutually exclusive. They simply provide alternative justifications for the validity of certain equations used to analyze the performance of real world systems. There is no reason why both justifications cannot be correct. Observational stochastics does offer practitioners a number of substantial advantages that have been detailed on the pages of this book. On the other hand, traditional stochastic modeling is a more mature discipline that has been applied successfully to many problems that are beyond the current scope of observational stochastics. It seems likely that both approaches will continue to yield useful and interesting results in the future.

References

Abramson, N.1970 "The ALOHA System." *AFIPS Conf. Proc.* 37, pp. 281-285.

Bertsekas, D.P. and J.N. Tsitsiklis. 2002. *Introduction to Probability*. Belmont, MA: Athena Scientific.

Birkhoff, G. 1931. "Proof of the Ergodic Theorem." *Proc. National Academy of Sciences* 17 (12), pp. 656–660.

Bolch, G., S. Greiner, H de Meer and K.S Trivedi. 2006. *Queuing Networks and Markov Chain*s. New York: John Wiley & Sons

Bruell, S.C. and G.F. Balbo. 1980. *Computational Algorithms for Closed Queuing Networks*. New York: Elsevier North Holland.

Buzen, J.P. 1971. *Queuing Network Models of Multiprogramming*. PhD dissertation, Harvard University. Also ESD-TR-71-345 (www.dtic.mil/dtic/tr/fulltext/u2/731575.pdf), Republished 1980. New York: Garland Publishing.

Buzen, J.P. 1973. "Computational Algorithms for Closed Queuing Networks with Exponential Servers." *Comm. ACM* 15 (9): pp. 527-531.

Buzen, J.P. and P.S. Goldberg. 1974. "Guidelines for the Use of Infinite Source Queuing Models in the Analysis of Computer System Performance." *AFIPS Conf. Proc.* 43, pp. 371-374.

Buzen, J.P. 1976a. "Fundamental Laws of Computer System Performance." *Proc. ACM Sigmetrics/IFIP WG 7.3 Symposium on Computer Performance Modeling, Measurement and Evaluation*, pp. 200-210. Republished as "Fundamental Operational Laws of System Computer Performance." *Acta Informatica* 7 (2): pp. 167-182.

Buzen, J.P. 1976b. "Operational Analysis: The Key to the New Generation of Performance Evaluation Tools." *Proc. IEEE COMPCON 76*: pp. 166-171.

Buzen, J.P., R.P. Goldberg, A.M Langer, E. Lentz, H.S Schwenk, D.A. Sheetz and A. Shum. 1978. "BEST/1: Design of a Tool for Computer System Capacity Planning." *AFIPS Conf. Proc.* 47, pp. 447-455.

Buzen, J.P. 2011. "The Rationale for Shaped Simulation." *Proc. TMS/DEVS '11*, pp. 83-88.

Buzen, J.P. 2012. "Computation, Uncertainty and Risk." *The Computer Journal* 55 (7): pp. 837-847.

Casale, G., M. Gribaudo and G. Serazzi. 2010. "Tools for Performance Evaluation of Computer Systems: Historical Evolution and Perspectives." *Proc. PERFORM 2010*, pp. 24-37.

Cox, D.R. 1955. "A Use of Complex Probabilities in the Theory of Stochastic Processes." Mathematical Proc. of the Cambridge Philosophical Society 51 (2): pp. 313-319.

Denning, P.J. and J.P. Buzen. 1978. "The Operational Analysis of Queuing Network Models." *ACM Computing Surveys* 10 (3): pp. 225-261.

Denning, P.J. and C.H. Martell, 2015. *Great Principles of Computing*. Cambridge, MA: MIT Press.

El-Taha, M. and S. Stidham, Jr. 1999. *Sample-Path Analysis of Queuing Systems*. Boston: Kluwer.

Erlang, A.K. 1917. "Solution of Some Problems in the Theory of Probabilities of Significance in Automatic Telephone Exchanges." Post Office Electrical Engineers' Journal 10: pp. 189-197

Ferrari, D., G. Serazzi and A. Zeigner. 1983. *Measurement and Tuning of Computer Systems*. Englewood Cliffs, NJ: Prentice-Hall.

Gelbaum, B.R. and J. M. H. Olmsted. 1964. *Counterexamples in Analysis*. San Francisco: Holden-Day.

Gelenbe, E and I. Mitrani. 2010. *Analysis and Synthesis of Computer Systems*. 2nd ed. London: Imperial College Press.

Gordon, W.J and G.F. Newell. 1967. "Cyclic Queuing Systems with Restricted Queue Lengths." Operations Research 15 (2): pp. 266-277.

Harchol-Balter, M. 2013. *Performance Modeling and Design of Computer Systems*. Cambridge: Cambridge University Press.

Harrison, P.G and N.M. Patel. 1993. *Performance Modeling of Communication Networks and Computer Architectures.* Workingham, UK: Addison-Wesley

Jackson, J.R. 1963. "Jobshop-like Queuing Systems." *Management Science* 10 (1): pp. 131-142.

Jain, R. 1991. *The Art of Computer Systems Performance Analysis.* New York: John Wiley & Sons.

Kleinrock, L. 1962. *Message Delay in Communication Nets with Storage*. PhD dissertation, MIT. Republished 1964. *Communication Nets: Stochastic Message Flow and Delay.* New York: McGraw-Hill.

Kobayashi, H., B.L. Mark and W. Turin. 2012. *Probability, Random Processes, and Statistical Analysis*. Cambridge: Cambridge University Press.

Kolmogorov, A.N. 1933. *Foundations of the Theory of Probability*. Berlin: Springer. Translation by N. Morrison. 1956. New York: Chelsea Publishing Co.

Lazowska, E.D., J. Zahorjan, G.S. Graham and K.C. Sevcik. 1984. *Quantitative System Performance*. Englewood Cliffs, NJ: Prentice Hall.

Little, J.D.C. 1961. "A Proof for the Queuing Formula: L = λW." *Operations Research* 9 (3): pp. 383-387.

Little, J.D.C. 2011. "Little's Law as viewed on its 50th Anniversary." *Operations Research* 59 (3): pp. 536-549.

Mealy, G.H. 1955. "A Method for Synthesizing Sequential Circuits." *Bell Systems Technical Journal* 34 (5): pp. 1045-1079.

Menascé, D.A., VA.F. Almeida and L.W. Dowdy. 1994. *Capacity Planning and Performance Modeling*. Englewood Cliffs, NJ: Prentice Hall

Menascé, D.A. and VA.F. Almeida. 2001. *Capacity Planning for Web Services*. Upper Saddle River, NJ: Prentice Hall

Metcalfe, R.M. 1973. *Packet Communication*. PhD dissertation, Harvard University. Republished 1996. San Jose: Peer-to-Peer Communications.

Metcalfe, R.M and L. Shustek. 2006. *Oral History of Robert Metcalfe*, CHM Ref: X3819.2007, Mountain View: Computer History Museum Publications.

Poincaré, H. 1905. *Science and Hypothesis*. London: Walter Scott Publishing.

Severance, C. 2014. "Len Kleinrock: The First Two Packets on the Internet." *Computer* 47 (3), pp. 10-11.

Stewart, W.J. 2009. *Probability, Markov Chains, Queues and Simulation*. Princeton: Princeton University Press.

Suri, R. 1983. "Robustness of Queuing Network Formulas." *Journal ACM* 30 (3): pp. 564-594.

Taleb, N.N. 2007. *The Black Swan: The Impact of the Highly Improbable*, New York: Random House

Index

www.ingramcontent.com/pod-product-compliance
Lightning Source LLC
Chambersburg PA
CBHW051852170526
45168CB00001B/76